1986

# A Layman's Introduction
## to Robotics

# A Layman's Introduction to Robotics

**Derek Kelly**

PETROCELLI BOOKS
*Princeton, New Jersey*

**Library of Congress Cataloging in Publication Data**

Kelly, Derek.
A layman's introduction to robotics.

Bibliography: p.
Includes indexes.
1. Robotics—Popular works.   I. Title.
TJ211.15.K45     1985       629.8'92          85-6591
ISBN 0-89433-265-1

To

MORGAN

A Child of the Future

and to

MARY CARLA

# Contents

iv

Contents

# Contents

# Contents

# Contents

# Acknowledgements

Almost every paragraph or sentence in this book could be expanded into a chapter or book. To keep a sense of perspective and directedness, compressions and omissions had to be made. Each of the separate chapters could be itself the subject for a book, and in some cases—for example, the chapter on history—several dozen volumes would be required to do justice to the historical background of robotics.

Though this book is short in relation to its subject, it is hoped that its brevity will present the ideas in a way that is more understandable for the general reader than the many technical and scholarly works upon which it is based. I am indebted to all those who have written in the area of robotics. The most helpful were David Heisermann, some of whose books may become classics in the field; Ludwig Von Bertalanffy, inventor of General Systems Theory; and Mr. Robotics himself, Isaac Asimov. Additional sources are to be found in the bibliographies.

I am indebted to friends, colleagues and other specialists in robotics who have helped me, in one way or another, to understand the theoretical and technological features of robotics. Among these are Doug Stowell, a former colleague; The Denver Robotics Club; Myron China and Floyd Inman.

# A Layman's Introduction to Robotics

As any writer would understand, I am grateful to those few friends that I have left after periods of extended neglect and inaccessibility, in particular, Mr. Charles Green.

Special thanks go to Dr. Steve Andriole whose comments and suggestions have much improved the overall balance of the discussions in this book.

One person to whom this book is dedicated is now only a young child. To her—and others like her—belongs the future, including the future that will be shaped by robotics.

It is an irony of life that most of us have it all backwards. We normally think that our elders are in fact our elders, and that youngsters are in fact youngsters, when the opposite is the case. Those of us who are "old" are in fact young in relation to the general culture since we started off with much less knowledge than do those who are born today or tomorrow. A newly-born child inherits the whole of existing culture, the whole ambience of its technology, and begins from that day onwards to add to that culture and knowledge. Thus, the youngest among us are in fact the oldest in terms of cultural inheritance, and those of us who are older in fact get younger and younger, in relation to the culture and its technology, as we get older. If we wish to stay young, then we must listen to emerging interests and participate in the emerging technologies of the day.

Though the difference between our ages is over a generation, I gladly dedicate this book, in part, to my cultural elder—Morgan.

The Author and Publisher wish to thank the following firms for permission to reproduce photographs in the plate signature between chapters 4 and 5: 21st Century Robotics, Radio Shack, Heath Company, RB Robot Corporation, Harvard Associates, Otis Observatory, Virtual Devices, Inc., Robotronix, Inc., Arctec, Inc., Prab Robotics, Inc., and Unimation, Inc.

# Foreword

In 1976, the first personal computer conference was held in Albuquerque, New Mexico. Eight years later, in 1984, the world's first personal robotics conference was again held in Albuquerque. The dominant attitude of those who were not involved with personal computers in 1976 was close to derision. Their principal question was, "But what can you do with a personal computer? Of what use is it?" Eight years later, personal computers are a multibillion dollar-a-year industry and no one is asking those questions anymore.

Today, the skeptics have renewed their questioning, but this time it is directed at personal robotics. "But what can a personal robot *do*?" they are asking, while implicitly suggesting that they cannot do anything now (which is close to the truth) and that they never will be able to do anything worthwhile in the future (which is very far from the truth). The truth of the matter is that the robotic industry generally, and personal robotics in particular, will become multibillion dollar industries by the 1990's. You can either be one of those who profits from this growth industry, or you may prefer to be an ostrich.

The intention of this book is to provide a generally educated but nontechnical reader with a continuously useable book that is not only informative, but also thoughtful, provocative, and practical.

The subject matter is robotics. At the present time, the technology of robotics is about at the stage where personal (or micro) computers were in 1976 or 1977; that is, robots are just beginning to enter the market as practical, labor-saving devices. I expect, along with many others, that the robotics revolution will be just as pervasive as the computer revolution, and that in a few short years, knowledge of robotics will be as necessary for economic survival as a knowledge of computers is today.

This book is an attempt to describe in a general way some central features of robotics, which is one of the principal players in the drama of the "emerging" post-industrial society to which we are all witnesses.[1] This emerging society and economic system is based on *information* (facts, knowledge, data processing) technologies that augment our mental powers using tools like computers, and on the emergence of *space* as a central feature of our visions of the future. Robots are the chief means whereby the information society will be realized. Robots are the pathway to the automated factories of tomorrow, to the use and development of information as a way of life, and to the exploration and economic exploitation of outer space.

Robots and robotics are relatively new phenomena to the general populace. Most everyone has heard of or seen or read something about robots, but robots are still regarded

[1]In his book, *2081, A Hopeful View of the Human Future*, Gerard O'Neill sees five driving industries in the year 2081 (p. 27): computers, automation, space colonies, energy, and communication. This book deals with one of those fields, robotics, and of its relations to these and other industries.

as something "strange out there." Here we wish to deal with robots in context, to dig out some of their salient features and to describe their implications for human life in general.

First, we need to see robotics in a context. Robotics does not stand alone, but is supported by many other disciplines and professions and fields of interest. Robotics is supported by mathematical logic, electrical engineering, telecommunication specialists, mathematicians, economists, and mechanical engineering. Similarly, it is bound up with management science, operations research (applied mathematics), systems theory and analysis, and cybernetics. Robotics is in fact the tip of an iceberg composed of disciplines like artificial intelligence (AI), computer programming, aeronautics and astronautics, solid-state physics, the psychology of perception, and many others. Robotics is relevant to and can make use of most existing industries and technologies—in new ways. Table 1 lists various fields of knowledge on the left column, with robotics and other industries in the headings. It shows

Table 1. Comparison of industries.

| Industry | Robots | Computer | Telecomm. | Production | Space |
|---|---|---|---|---|---|
| Math | x | x | x | x | x |
| Electronics | x | x | x | x | x |
| Systems | x | x | x | x | x |
| AI | x | x | x | x | x |
| Chips | x | x | x | x | x |
| Management | x | x | x | x | |
| OR | x | x | x | x | x |
| Programming | x | x | x | x | x |
| Aeronautics | x | | x | | x |
| Perception | x | x | x | | |

the relevance of the fields of knowledge to robotics and related industries.

A final purpose is to present not only theory but also practicalities, definite details that can help people to orient themselves better to the "new" world in which we all now live and work and play. The prevalence and productivity of robots in manufacturing and other industries and the imminent democratization of robots (brought about by the personal robot), and thus their affordability to just about anyone, is a phenomenon that will affect every part of our society, from its work habits and economy to the dreams and aspirations of each member of the society.

As a consequence of its widespread effects, there is a need to see all these details about robotics in terms of a framework of interpretation. The framework chosen here is to deal with robotics not simply as a technological or economic phenomenon, but as a general cultural phenomenon. When approached as a cultural phenomenon, robots are seen as having an effect on our past and future work and lifestyles.

My perspective to the subject of robotics is a general one. Though trained at the doctoral level in the ideas and theories that underlie robots and robotics, I also have practical business experience. Though I greatly respect and admire theoreticians of many stripes, I am nonetheless thoroughly convinced that theory without practical experience is as empty as a practical, bottom-line business orientation is often blind. My hope is that my perspective and treatment of robotics will be thought-provoking and of continuing interest to the reader.

# Introduction

The purpose of this introduction is to describe the context, situation, and state of affairs that makes robotics a topic worthy of the reader's interest and attention.

In his 1980 book, *The Third Wave*, Alvin Toffler, a well-known "futurist," mentions robots only once (in the course of a very long book about the future). This is a curious neglect in the work of a man who has made the "future" famous. Unfortunately, many of the books by Toffler and others dealing with the future are pedestrian works, shackled to the past and to the experts who spend their time "looking backward," so this well-known futurist completely missed the whole phenomenon of robotics which only a few years later is spreading throughout the world.

In 1980, there were only about 1,500 robots at work in American industry. By 1985, the expectation is that there will be 5,000 to 10,000. In 1984, two significant events occurred which foretell a rapid development of robotics in the years ahead. First, the First International Personal Robot Congress and Exposition was held in Albuquerque, New Mexico, where the MITS Altair (the first microcomputer) was born. The second event was the Robots 8 Exposition and Conference of industrial robotics sponsored by the Society of Manufacturing Engineers (SME). That year also saw the beginning of the widespread coverage in many media of robots. Robots have been the topic of talk shows, television sitcoms, and numerous educational and scientific programs in the media. Almost on a daily basis, one can expect to see and read an article on robotics in the local paper. There is thus apparently a growing public awareness of and interest in robotics.

The robotics industry is on the verge of breaking wide open. Where the industry is headed and what its development foretells for our future is something that needs to be dealt with.

## The Problem Situation

First, we need to understand the context in industrial society within which robotics has arisen. What is the phenomenon of robots and robotics? What are robots? What are their roles? Are they as terrible or as wonderful as people say they are?

To answer these questions, we need a context for understanding the role of robotics in the present age. In what follows, three perspectives on the context of robotics are presented. First, we will consider the nature of the post-industrial society where 60 percent of the economy is service-oriented, the remainder being in manufacturing. Second, we will consider the nature of manufacturing and its relation to robotics. Third, we will describe robotics as a phenomenon in and of itself.

# Introduction

## Robotics in the Information Society

Daniel Bell[2] tells us that we live in a post-industrial information society based on computers and telecommunications. In his view, industrial society was based on energy and on products related to energy; that is, to coal, steam, oil, gas, and other lesser used forms of energy like solar, wind, and hydro. The economy of the industrial society gave us locomotion and consumption as its two main products. For locomotion it gave us energy-extensive (and increasingly expensive) trains and automobiles and airplanes. It also gave us artificial fabrics with which to cover ourselves, many useful household items based on oil-extracted products, and numerous labor-saving devices like refrigerators, washer-dryers, toasters and other energy-guzzling appliances.

One of the features of the energy-based economy was the whole economically wasteful phenomenon of centralized offices, with workers spending exorbitant amounts of money on gas and parking to get to work each day, and corporations spending exorbitant amounts on property and other taxes to provide their workers with suitable working conditions. One of the central features of the information revolution is that it diminishes the need for central offices—people can work from home terminals—and for all the unprofitable things that have to be done to maintain centralized offices: suburbs where urban workers can sleep, cars so they can get to work, parking spaces where their cars can sit all day doing nothing useful, etc.

With the advent of the portable telephone, an individual can connect a home terminal to a "home" computer through a local telephone call from just about anywhere in the world.

---

[2]See Daniel Bell, "Social Frameworks of the Information Society," in Michael L. Dertouzos and Joel Moses, Eds., *The Computer Age—A Twenty Year View*, Cambridge: MIT Press, 1979, 1980, pp. 163–178.

# A Layman's Introduction to Robotics

In the next two or three years, two technological advance-
ments will enable people to work with information systems
from anywhere in the world, whether there are telephones
there or not. One advancement is the proliferation of communication satellites which will eliminate the need for telephone lines as a medium of communication. The other
advancement is the development of devices that will enable
people to access computers and other information systems
through satellites rather than through telephone lines. We
can thus expect that by the end of the current decade, individual members of an information group will maintain their
cohesiveness even if scattered over the world. One individual
who may prefer to work from atop a lonely mountain can
just direct his satellite communication device in the appropriate direction, then access his "home" computer through
his battery-powered terminal from wherever he happens to
be.

In contrast to the industrial economy (see Table 2), the
post-industrial economy is based on completely different
products and effects. The information society is appropriately
based on information, on the production, distribution, and
accessibility of data bases, information sources, reports,
instant (tele-)communications, and powerful processing
capabilities. Energy-related products are "heavy" items; they

Table 2. Industrial & post industrial societies

| Category | Industrial Society | Post-Industrial Society |
|----------|-------------------|------------------------|
| Basis | Energy | Information |
| Economy | Consumption | Learning |
| Quantity | Bigness | Littleness |
| Motive | Grandiosity | Ephemeralization |
| Work | Drudgery | Pleasant |
| Leisure | Escape from work | More Work |
| Location | Work at office | Work at Home |

are tangible, physical, practical and manageable. Information, on the other hand, is weightless, intangible, mental. To produce heavy products, people worked in foundries as welders, painters, stock clerks. The work in such cases is wearying and unfulfilling. Workers spend their time waiting to "get off" work and to relax doing something other than work. After a day of back-breaking labor such as is required of workers in the "heavy" industries, no one has much energy or interest left to do other things, and thus such people dig into their beer and slink down in front of the passive leisure supplied by TV.

By contrast, the information society is based on a form of work which is intrinsically different from the work in "heavy" industry. The offices, workplaces, and work styles of information industries are not of drudgery but of challenge and excitement. As robots take over more and more of the "productive" functions—the manufacturing—of the society, people will be free to work in the areas where people work best: in service industries, in activities where understanding, empathy, interpretation and judgement are required.

We are in need of and on the verge of an economic and technological revolution. Using people to provide the "dirty work" needed to produce the consumer items needed in a consumer-oriented society has become increasingly costly and inefficient. Besides that, people should have some other means of making a living other than in drudge work. The result is that the information industry is now deeply ingrained in the society. The ownership and use of computers and other information processing devices is growing.[3] IBM and AT&T,

---

[3]Apple computers (*Wall Street Journal* of 6/30/83) just passed one million Apple IIs. By 1987, industry experts predict that there will be 60 to 70 million personal computers in the field. By that time, computers will have become as common as cars. A similar phenomenon can be expected with respect to robots around the year 2000. Computers are expected to be as common—and as necessary—as cars and toasters in the near future. By the turn of the century, we can expect the same state of affairs to prevail with respect to robots.

the country's two largest information corporations, have dedicated billions to providing us with machines and software for processing information, and other large corporations are dropping energy-related investments for high-tech investments, among which information processing machines such as computers are central.

One of the clearest differences is between the view of work and leisure in industrial and post-industrial societies. In an industrial society, work is something to escape from; in information economies, work is something that flows imperceptibly into leisure. In the high-tech computer economy, for example, programmers buy personal computers so they can "work" at home doing the same things they do at work.

Robots are a natural outgrowth of computers. Not only are many high-tech people getting computers to use at home, but they are also getting personal and simple industrial robots to connect to their personal computers. Computers are mental machines. Robots need those mental machines, but they have the added dimension in being able to *do* as well as compute.

## Robotics in the Context of Production

Now, we need a second context for understanding the development of robotics. This context is provided by the manufacturing industries which are making the most rapid and widespread uses of robots in the workplace.

In 1981, manufacturers in this country realized most forcefully that Japanese manufacturers were more efficient and productive than their American counterparts, and that unless changes were made, American products would become uncompetitive on the world market.[4] The Japanese were (are)

---

[4]See Jean Jacques Servan-Schreiber, *The World Challenge*, N.Y.: Simon and Schuster, 1981. See also Gene Bylinsky, "The Race to the Automatic Factory," *Fortune*, vol. 107, 4, Feb. 21, 1983, pp. 52–56.

producing better products at less cost than are Americans because, in part, they are using robots. (By "better" we mean more reliable and more trouble-free with fewer recalls. By less costly, we mean not that the Japanese work for less, but that robots that cost considerably less than human labor do a much better job.)

The claim that the Japanese are more productive than Americans is not related only to manufacturing. In point of fact, all economic and other activities (such as painting) are productive activities and involve the same steps as are involved in manufacturing: determining the internal or external need for a product (e.g., a service), planning and designing a way to meet the need and produce the product, designing the product, making it, and delivering it. So the idea of greater Japanese productivity is not related only to "manufacturing" productivity, but also the service-related sides of the economy as well.

Moreover, the difference does not lie only in the way manufacturing is done in this country but also on who does it and how it is done. Effective, productive work of any kind requires some sort of stability in the workplace and reliability in the worker. Part of the problem for us caused by Japanese productivity is that there is little stability or security in the workplace as businesses hire and reduce their forces periodically and consistently.[5] Machines, however, do provide greater reliability and capability than people, and given robots that can do the manual and intellectual activities of humans, we will have reliable and productive workers.

Simply as a matter of business, profit is achieved by making the result worth more than the effort that went into it. As a matter of a business decision, whatever helps to reduce the cost and improve the service related to a product

---

[5]Japanese society is based on "collectivities" and secure employment for life for all. In the USA, individuality is one of the central bases. These are two different mindsets; methods of work productive for one are not necessarily so for the other.

A Layman's Introduction to Robotics

must be done. We know that cars made on Wednesday are usually the best cars; statistically, such cars have fewer problems. Since robots do not have to spend the last two days of every week looking forward to the weekend, and certainly are not likely to show up for work on Monday with a hangover, the exorbitant costs of recalling and fixing the Monday and Friday "lemons" can be saved through their use.

There has been a noticeable decrease in the reliability and continuity of the workforce in the industries. Pride in work, doing one's best, and industrious work habits are not the norm in the workplace but only pleas on the part of management. Shoddy performance by industrial workers is a major headache for the economy. The reason is not always necessarily because the person is lazy, though that does happen, even in high-tech companies, but because of a lack of education—at a time when more and more jobs are technologically sophisticated, more and more workers are technologically illiterate[6] and thus unprepared for anything else. Both because of the unsatisfactory nature of manufacturing jobs and because it costs too much otherwise, the race is on to automate the factory, to use intelligent machines to do what humans normally do, and to use human beings where they work best—controlling, monitoring, guiding, and directing the work of machines.

Manufacturers of products and producers of any product or service want to be effective and to realize some return on their investment of time and money. These industries aim to satisfy the needs for and demands of their customers while at the same time keeping their expenses lower than their revenues so as to yield a profit.

One can make a profit in two basic ways:

By liberal expenditures and multitudes of sales,

[6]See The National Commission on Excellence in Education report, *A Nation at Risk*, April 26, 1983. See the main section, . . . "History is not kind to idlers."

xxii

Table 3. Projections of robots in industry.

| Year | USA | $ | JAPAN | $ |
|------|-----|---|-------|---|
| 1980 | 1,500 | 100,000,000 | 77,000 | 3 billion |
| 1985 | 6,000 | 2 billion | 1,000,000 | 50 billion |
| 1990 | 150,000 | 20 billion | 5,000,000 | 200 billion |
| 1995 | 1,000,000 | 500 billion | 10,000,000 | 300 billion |
| 2000 | 2,000,000 | 100 billion | 20,000,000 | 50 billion |

By keeping tight control on expenditures and capturing a solid share of the market.

American productivity has up to now been based on the first, and Japanese productivity on the other. Given a land of abundant natural resources, the first alternative makes good sense. Given a situation of increasingly expensive and scarce resources, the second alternative makes sense.

One of the ways in which control has been exerted over production processes is through the use of computerized production control systems. Instead of the hopelessly inefficient and time-consuming calculations that were once done by hand, computerization allowed manufacturers to gain greater control over the investments they had to make in inventory, while at the same time increasing the service provided to their customers. With machines that can do "human" jobs, there is the possibility of exerting more control. Robots get the job done for less cost, and it is done better by workers who don't get sick, tired, or drunk.

Increasingly, robots, and in particular industrial robots, are providing reliability and continuity in the workplace. Table 3 shows the expected number of robots and expenditures on them in dollars for 1980 to 2000.[7]

[7]The figures for 1980 and 1985 are from standard industry sources. See, for example, "Analysts predict a breakout for robotics in 1984" in *Computerworld*, March 5, 1984, p. 89. The figures for the remaining years are based on a simple model. The figures

# A Layman's Introduction to Robotics

In industry alone, robot use is expected to grow astronomically over the next few years. To understand robotics, we need to see it as rooted in the transition to an information-based economy with automated factories.

## The Context of Robotics

Though human beings have been buying labor-saving devices to "replace" human labor for ages, there is no longer any great fear in knowing that a refrigerator, or a washer, or a toaster, or an oven not only "replaces" human beings but does the job a lot better than they ever could.

But when it comes to robots (automata) there are now fear and anguished expectations on the part of many people.[8] If robots can be made (perhaps only eventually, though not now) that can see, touch, feel, manipulate, get around, learn, and do many jobs that human beings do now, only do it reliably, continuously, without any backtalk, without complaints, without any need for paychecks or benefits (but certainly maintenance) or retirement funds, . . . then of what use are humans? What is to be done to or done by people whose jobs have been taken over by robots? According to the report, *A Nation at Risk*, there is nowhere for these people to turn. In fact, the fears of unemployment and displacement are legitimate fears, for *the technologically illiterate are doomed*

for 1980 are taken as the "level" of a trend model. The figures for 1985 are taken as a "trend". A straight line is drawn from level to trend and extrapolated to the remaining years.

[8]This fear is long-standing. In the mid-1600s, when Blaise Pascal tried to commercialize a machine that could add, subtract and multiply (e.g., like a supermarket register), it was a commercial failure because people expressed fears about the unemployment such a machine would cause, how service and repair would be provided in the event that Pascal, the only one who knew how to build and repair such machines, was no longer around. Similar fears are expressed about robots (and computers) today. Such a similarity of response to a novel machine leads me to suspect that arguments against robots and other machines are not based so much on fear of machines so much as fear of change, fear of the new, fear of the uncomfortable.

*to economic extinction* in the information society.[9] Since the new jobs created by information are always technologically demanding, workers without training in the new technology have no chance for a job without retraining. The net effect is that in the transition from a "heavy" industrial base to a "light" electronic industrial base, the economy has to be able to support a whole generation of technologically ill-equipped workers. Where can the ordinary, untechnical, computer-illiterate, robot-ignorant person turn to make a living when replaced by robots? The answer, not surprisingly, is to *use* robots, not to *make* them. (The Japanese have factories where robots manufacture other robots. The jobs are not going to be in manufacturing on the shop floor for very much longer.)

Robots seen as replacing human beings and their jobs yields one view of automata, while the view of robots as helpers and expanders of human powers yields yet another view.[10]

At the present time, there is widespread fear of robots. Labor unions, in particular, are afraid of the steady erosion of jobs by robots. On the other hand, there are many people who do not fear robots, but welcome them as a way whereby we can further develop the human species. In this book, attempts will be made to show that the fear of robots is unreasonable, and that people should in fact be welcoming robots into their lives. So what is this new technology and industry that promises to fulfill the computer and that is expected to revolutionize the manufacturing industries by the turn of the century? Some of the principal questions for which we intend to provide some answers are:

[9]"Fear of loss of jobs is at the root of workers' objections to robotics." In "People Make Robots Work," by Fred K. Foulkes and Jeffrey L. Hursch, *Harvard Business Review*, Vol. 62, No. 1, Jan.-Feb., 1984, p. 94.

[10]A pioneer in constructing calculating machines, which are the forerunners of computers and robots, Leibniz (1646–1716) once said: "It is unworthy of excellent men to lose hours like slaves in the labor of calculations which could safely be relegated to anyone else if machines were used." (See Jerry M. Rosenberg, *The Computer Prophets*, p. 49; see bibliography for details.)

What is a robot and what is robotics?

What is the historical origin of robotics? Where did the idea come from? How is it related to people and events in human history?

What machines and other technology are used in robotics?

What sorts of software or procedures are available? What controls and applications are available or foreseen? What, in short, can robots *do*?

What careers can one prepare oneself for so as to fit into one or another area of the field of robotics? Where are the jobs now? Where will they be in the future?

What are the implications of robotics for work for the general well-being of the society and the people within it? Do we have anything to fear in robots? Is there anything hopeful about the appearance of robotics on the scene?

What does the future portend for robotics? What are the trends and possibilities?

Last but not least, where can one find more information, other perspectives, and additional details on robots and robotics?

## Who is the Reader?

What is the goal of this book, and to whom is it addressed?

My purpose in writing this book is to organize and describe knowledge and ideas about what I believe to be a central industry in the society and economic system that will be around for many years to come, and to relate this industry to the wider society so that other people may gain an understanding and sense of curiosity about it.

# Introduction

Robotics is the second of three industrial waves that will together be the most pervasive features of the 21st century—just 15 years hence. The first wave was begun by computers. The third wave will be the permanent human habitation of space. The second wave, robotics, is the connecting link between the two.

If you, the reader, are a professional in robotics then this book is directed to you. What you know about your own field may be compared with the views presented here, and also you may gain from seeing connections drawn by the author, connections which indicate a perspective, a way of perceiving and conceiving of robotics.

If you are a student preparing for a career, then you may find guidance in the overview of robotics that is provided here, and also from the description of the careers that are and will be available in this field. If you are already working in the computer field, all you need do to get into robotics is simply express an interest. If you are not in the computing field at present, this book will provide you with a basic orientation and preparation for the robotics field.

If you are a general lay reader, this book will provide you with both general and specific information about the industry that is shaping the society in which we live.

# What is Robotics? 1

## Introduction

A robot is both (1) an automaton and (2) intelligent. It is an automaton in that it is a machine that can control, in some degree or other, its own actions. It is a general manipulator in the sense that it is a machine built with the capacity to do many different things, perform many different intelligent actions. Robotics, on the other hand, is a field of interest that combines theory and application, ideas with actual practical machines. Robotics is a field of activity that combines both knowledge of the theoretical disciplines which underlie robots, as well as experience or interest in the practical applicability and actual implementation of applications using robots.

Robotics is thus both a noun and a verb, a thing (a body of knowledge) and an activity (a doing). When used as a

1

noun, as a name for a type of device, the term robot refers to automatic devices, automatons, that is, self-moving, self-acting or self-controlling machines.

The ability to self-move is akin to human abilities. Humans often distinguish themselves from machines precisely in terms of the ability to move oneself. What this means is that the ability to move oneself means a lot more than it sounds.

One of the fundamental distinctions in terms of which we understand ourselves is that between the voluntary and the involuntary. Involuntary actions are those which happen without conscious control. Voluntary actions, on the other hand, involve intentions and thus conscious awareness. Both voluntary and involuntary actions happen according to "laws," i.e., according to the laws of nature governing the behavior of physical beings. However, the laws that apply in each case are different.

Neither the rock nor the heavy rain that jarred it loose from the hilltop are voluntary—they are involuntary. On the other hand, human beings can perform voluntary actions—actions which result from a sequence of phases that goes: (1) awareness of problem, (2) hesitation and deliberation over alternative solutions, (3) consideration of the reasons and motives for and against possible solutions, (4) selection of a solution, and (5) moving the body to carry out the decision.

Human beings are also subject to many involuntary actions. We usually reserve the term "accident" to refer to such occasions: stubbing one's toe is generally not a voluntary but an involuntary action, as is having an automobile accident.

Robots can be considered as artificial beings that have the capacity for low-level awareness and thus voluntary activity in pursuit of internally or externally defined goals.

When used as a verb, "robotize" has the implication of automating, programming, controlling. A human being, allegedly, can decide (1) what principles to use in deciding whether to move or not and (2) can delay or decide not to move; automatons are "told" (programmed) to move under such and such conditions and are not given the right—or do not have the technological capability—of "making up their own minds". Not having this right or ability is equivalent to being a slave, a servo-mechanism, a controllee at the service of some other creature, namely human beings.

When "robot" is treated as a verb, as a flowing, moving, effectuating, manipulating term, it resembles a vector that moves from being a noun (a point) to being a direction, a drive towards something, as depicted in Figure 1. Thus, as

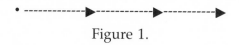

Figure 1.

a way of talking about past, present and future action, it takes on an additional sense. The first sense was a capability, a raw capacity of awareness and low-level conscious intentions; the second sense is now a capability related to action, to movement, to performance.

So, we can say that the term robot refers to a machine, an artificial organic or inorganic mechanism capable of some degree of awareness using human-conscious capacities as a model, and which is capable of manipulations and actions that resemble or are modelled on human manipulation and action.

This brings us to a third sense of the term robot, as both a noun and a verb, as both a "conscious" something or other, and as something that can duplicate an increasing number of human capabilities.

When used as both a noun and a verb, that is, as a description of what something *is*, or of what someone is like, to be robot-like is equivalent to being uniform, equitable, pleasant, even-tempered, constant, reliable, consistent (quite the opposite of human beings, please note).

One human characteristic is to be unreliable (exceptions

3

Table 4. Robots and humans.

| Being | Consciousness | Action |
|---|---|---|
| Human model | High level | Multiactive |
| Robots today | Low level | Few actions |
| Robots in near future | Low level | Few actions |
| Robots in far future | Higher than humans | More than humans? |

are referred to as people of "great character"), spontaneous and unpredictable, nonuniform except in totalitarian or dictatorial states, unpleasant, odd-tempered, inconstant and inconsistent. To be a robot is, then, to be (1) at odds with most of human characteristics, for only exceptional people (without making any judgment here as to the value or disvalue of these characteristics) are robot-like, and (2) it consists of an ability to do things that human beings can do— but not consistently enough—and to do them continuously and constantly. When a robot is programmed to do such and such, it does not take coffee breaks, it does not slack off, it just continues to do what it was programmed to do until an external agent (like a human being) or an internal control mechanism tells it to do otherwise.

We create machines that are able to do things we'd like to do but can't because we get tired, or our minds wander, and so on. A robot is a being-actor in an evolving sense: robots are now at a low point in their conscious and action capabilities, and similarly are at a low level in their manipulation capabilities. But the aim behind the development of robots is the development of artificial beings with conscious and active capabilities much like, and even greater than, those of human beings. Thus the term "robot" as a being-actor is an evolving term: robots can be expected to evolve to greater and greater powers.

4

## Robots and Automatons

Robots are automatons, but not all automatons are robots. In general, automatons are self-acting, self-propelling, self-governing, self-regulating servo-mechanisms. They are mechanical or electro-mechanical devices created with these characteristics.

The reader should be aware that automaton does not mean *autonomous*, though this is subject to dispute.[11] To be autonomous, a characteristic reserved for humans (or higher) beings, is to be *answerable* ultimately to oneself, to be in the final analysis the one who "decides" one's own course in all things. *Auto* means "self"; *nomos* means law; thus, autonomous means literally "law unto oneself".

An automaton is not autonomous in this human sense. An automaton is a self-moving device but not a law-unto-itself device. No matter how complex and capable automatons may become, no automaton can be autonomous, by definition. If an automaton is given, or develops for itself[12] autonomous capabilities, then the device would *ipso facto* cease being an automaton, and would have a right to claim for itself rights of civil, economic, and legal natures as do all autonomous beings. One example of autonomy in the human sense comes to us from Sweden. In Swedish society, a child of five years can "decide" which of two parents it wishes to live with, and can, moreover, renounce a parent as inadequate. The child is given legal and civil rights normally reserved for "older" autonomous beings. A child of five may not have the same capacity of judgement as an older person, but its level of judgement is taken as sufficient to make the

---

[11]Robots are often spoken of as having degrees of "autonomy," thus this is not autonomy in the sense in which humans have it, but as an analogue of this human power.

[12]This is not farfetched, as one of the areas of current research in the field of artificial intelligence that underlies automatons and robots is the development of self-reproducing machines.

5

judgement that one or both parents are unfit to live with. Similarly, if a robot were to attain the mental and manipulative capabilities of a five-year-old, then it would have a basis for claiming certain rights "under the law". Whereas destruction of personal property by another is a crime against property, a crime against a personal robot may in the future be regarded as a crime against a "person".

## Development of the Term "Robot"

The term "robot" became popular as a result of the success achieved in 1923 by Karel Capek's play *R.U.R.* As used in the play, a robot was a zombie-like device, a mute but capable servant. The term "robotics" was invented by Isaac Asimov in 1942 to denote the study of and development of applications for robots.

With the advent of computers and artificial intelligence devices with the (cybernetic) capability of "self-control," the term robot has come to mean a device that can be programmed to perform many different tasks, some of them requiring artificial intelligent components. That is, a robot is capable of many "practical" or mechanical tasks, and it needs some intelligence to be able to perform these functions. But no matter how intelligent, the fact that they lack autonomy makes them slaves in Capek's sense—they can do anything, but they do it at the behest of humans.

## Intelligent Capabilities of Robots

"Intelligence" capabilities have been developed by the field of knowledge known broadly as artificial intelligence (AI). These capabilities depend on computer technologies for their actualization. The "manipulation" capabilities of robots are the work of electronics and mechanical technologies. At the present time, as we will see, the intelligence capabilities of robots, though quite remarkable, are still in the primitive stages of development. No robot can duplicate human powers of sight, hearing, dexterity, and thought. Not even androids, mechanical animals, are very far advanced. For example, no robot can recognize patterns very well, and no robot has the mechanical ability to climb stairs very easily. Just duplicating animal capabilities is hard enough; robots are not even close to going beyond human and animal capabilities.

There are two types of automatons:

single function, single program (e.g., a car);

multifunction, multiprogram manipulation devices.

"Automaton" can refer to both these types, but the term robot is limited to the multicapability version of automatons. However, we will consider both types of automatons in this part of the book inasmuch as the line between multi- and single-function automatons is hard to draw in real life, and many of the functions that can be performed by robots derive from single-function devices.

## Robotics as a Melding Technology

While both the information technologies and the manufacturing industries involve the development of new technologies and new techniques, the robot industry does not represent new theory, or new technology, or new techniques beyond those already used by information systems and production situations. The present "status" of robots and the growing public awareness of the so-called "robotic revolution" is not the result of any new technological or theoretical discoveries. The popularization of computers and the microcomputer revolution was caused by the development of the silicon chip, pure and simple. No equivalent development, up to this point, is involved in the robotic revolution.

As a melding of computer and electro-mechanical technologies, the robot is nothing new, but simply a continuation of the revolution begun by computers, by devices developed to augment the powers of the human mind (as distinct from all previous inventions that were designed to improve the powers of the human body, e.g., levers, motors, pulleys, and so on).

This melding indicates that the robot is the key to the transition from industrial to post-industrial society, for it combines both the residue of industrial society in the form of manipulation capabilities and is also a mechanical person with more than just manipulation capabilities. Industrial economy produced many wonderful physical machines; the information society produces mental machines. Robots are a key that combine both economies, both principles.[13]

What may be regarded as a revolutionary breakthrough in robots was the discovery or realization that while adding up numbers is something that a purely "mental" machine can do (e.g., a computer), really interesting intellectual and intelligent feats require that a "mind" have a "body"—that is, a set of powers for sensing its environment—and intelligent capabilities for interpreting the environment and of "deciding" how to respond to it. The robot really is revolutionary when it is seen that robots are the melding of "mental" and "physical" technologies. Dreaming up ways of artificially creating intelligence is one thing; creating the physical, technological way to actualize that intelligence is another thing altogether. So far, the electro-mechanical technologies involved in robots and androids are interesting more for promise than for accomplishment.

---

[13]See John Naisbitt, *Megatrends*, New York: Warner, Chapter 1 on the transition from an industrial to an information society.

## Is The Robot a New Species?

According to John G. Kemeny, President of Dartmouth College, and the inventor of the popular BASIC programming language, the robot is a new species of machine—a telemachine, a teleonomic machine. It is self-acting, self-moving, self-activating—it is a mechanical man. The robot is a new species of machine that combines the mechanical machine and the electro-computational machine, the clock and the computer, into a new combination.

The clock is the most perfect example of an automaton. Feed it energy and it moves of its own accord until its energy is depleted. The clock is a model for all prerobotic machines. It regularizes human behavior, and paved the way for the development of organized labor. Computers, on the other hand, are examples of mental machines.

Robots are a melding of both technologies—those of the body and those of the mind.

## Single-Function Robots

The aim of robotics is the development of an intelligent general manipulator. As part of the process of this development, there are many companies that are developing single pieces and functions, single bits of intelligence. In general the business side of robotics can be regarded as the *beta site* of AI, the testing ground of AI. The result is the proliferation of single function, single program devices with manipulatory abilities—but hardwired and limited. These are part of the multipurpose, and are not interesting as devices independent of multiprogram robots. For while the economic advantage of single-function robots may at present be quite great, a machine that can only paint or weld is a single-function machine not significant in and of itself as a robot. The majority of robots today are used in manufacturing; they are often bolted to the floor and do *one* thing very well.

Industry is using single-function robots increasingly in its operations. The automobile industry in particular is using robots to replace human workers in doing such things as painting, welding, and other activities that are dangerous, hard, or which require the repetition of a set movement over and over again, or which require constant precision, e.g., in assembling electronic devices.

Instead of being sad about this state of affairs, people should instead be glad. One of the sociological and psychological phenomena of industrial society is the extent to which it dehumanizes work, makes work a drudgery. To the extent to which robots take over the principal drudge work in society, that is the extent to which dehumanizing work will be taken out of the hands of human beings. That is something to look forward to. Unfortunately, many people are afraid about their jobs and the loss that entails for themselves, but this loss is in fact a gain as we will see later.

## Multifunction Robots

The Robotic Institute of America, a professional organization dedicated to the body of knowledge of robotics and to the development of professionals in the field, defines the robot as a *reprogrammable, multifunction, general manipulator*. Since single-function robots are special cases of these general manipulators, we can gain further understanding of robots by considering this definition of a robot at greater length.

### Reprogrammability

To be reprogrammable, a robot has to be programmable in the first place. To be programmable, the robot must have some way whereby procedures and instruction sets (programs) can be entered to the robot's "mind" (computer memory). A programmable robot is a robot that at the very least has the capabilities of computers, and primarily the programmable capability.

A program is a defined series of steps that produce a solution to a problem. That program must be stateable in the language of a computer, and must be stored in a computer's memory and recalled for use either directly, or indirectly, as in a subroutine, whenever desired.

A robot must be programmable to perform procedures or functions. This means that in addition to having a computer "mind" that is capable of storing and recalling lists of instructions (as a computer is able to do), it must also be able to *execute* those instructions. These instructions must be executable not on some CRT or video display screen, but in physical movements and manipulatory actions. So, in addition to having the general capabilities of computers, robots must also be able to perform. They must have electro-mechanical capabilities for doing what they are told to do.

Robots are also reprogrammable. This means that a robot is not a device that was "built to do one thing" (one program and one set of actions), but one built to do many different things, built as a general manipulator, and able to be programmed to do many different things. We are still not necessarily at the level of general manipulators. Industrial robots can be programmed to do different things, but most industrial robots are bolted to the floor and use a single capability, e.g., an arm, to perform different manipulations.

The first part of the idea of programmability and reprogrammability, namely the possession of a computer "mind," is fairly easy to accomplish for a robot. Since existing computers are programmable and reprogrammable for a wide variety of "symbol" manipulation devices, robots can simply adopt wholesale existing computer technology for use in their brains. The core of machines that compute is the silicon chip, many of which can be used by a robot. The one big problem is that even though our computers today are very powerful,

11

A Layman's Introduction to Robotics

the powers that they have are still not enough to provide us with intelligent robots.

To do the tasks we want robots to be able to do (see Chapters 3 and 5), robots will need great intelligence—and perhaps even the ability to duplicate and improve on human intelligence itself. One example may help to clarify this point. A Cray-1™ is one of the most powerful computers produced in this country today, costing about $10,000,000, and used where rapid and multiple computations are required at one time, as in weather forecasting or in tracking ballistic missiles. The power of a Cray-1 is built into the nose of the Cruise missile. That power is needed if the missile is to be able to distinguish between three identical buildings, say, standing side by side, one of which is the "real" target. Intelligent robots are going to need to make many such decisions and will need more power than is in a Cray-1 if they are to be really intelligent and self-contained. In all likelihood, we will need chips based not only on large-scale integration (LSI) that are now used throughout the industry, nor even the VLSI (very large-scale integration) chips that are currently being developed for the fifth generation of computers, but chips based on organic or atomic models.

But when it comes to the second part of the idea, namely the ability to be a general manipulator, the capability to accept any program and execute any physically possible action, that day is not even close. The physical technology is way behind the "symbolic" technology. Sure, computers can be programmed to add numbers, write letters, display wonderful graphics, and save the entire galaxy from ruin, but can a computer be programmed to *do* anything simple like sweeping my floors?[14] No. Robots can be programmed to follow a predefined pathway or to use a grid system to "map" out an environment, but the day when we will in fact have a general

[14]The RB Robot Corporation of Golden, Colorado, announced a vacuum sweeper attachment and program for their RBX5 personal robot for the Fall of 1983.

manipulator—a mechanical human being—is still far off in the future.

## Multifunctionality

Multifunctionality is a second idea or criterion for a robot. A multifunctional robot needs to have:

movement

maneuverability

manipulation tools—arm, wrist, hand

sensors

intelligence.

A general-purpose robot needs to be able to move itself around. It needs to be able to crawl through narrow openings and run up stairs. Unfortunately, these things are not possible yet. The physiology of robots is today very crude. The "body" in which robotic mentalities are stored is clumsy in most cases. There's something very nice about having two feet, two hands, and so on. Most robots have wheels for feet, an arm and a hand with limited capabilities, and the maneuvering capabilities of a bull in a china shop. Given this one required feature, we can say that we do not yet have general-purpose, multifunctional robots.

In addition, such robots need to have great maneuverability. For that they need a center of gravity and a "sense" of balance. They need arms and legs and sensing devices of great acuity. These need not be two-legged. Recently, a six-legged robot was announced that is able to walk through a stump-studded environment and can climb easy stairways. So, the day of the robot that can walk up and down stairs should not be too far behind.

Multifunctional robots also need to have ways to manipulate objects in their environment. They need arms, wrists,

13

shoulders, fingers, and so on. This part of the industry is piddling along with some interesting developments that we'll see when we take a look at robotic hardware.

Multifunctional robots also need to have many sensory capabilities. Sight, hearing, touch, taste, and smell sensors are available, though their powers vary. While military satellites can photograph any selected square foot of the earth with the resolution of a photograph taken just a dozen feet away from that square foot, the resolution powers of the photograph have not yet been merged with the resolution powers of sight or optical sensors. It should not be too difficult for technology to develop optical sensors capable of x-ray vision, microscopic and telescopic vision, and so on. These are well within the reach of the next 20 years.

There is also an economic reason to support multifunctionality. Manufacturers have found that the economics of robots is such that robots are far *cheaper* than humans. That being so, the more functions they can perform, the greater the productivity that can be derived from them (see Table 5).

Hearing has similar expectations. Auditory sensors capable of "hearing" down to say 1 or 2 cycles per second (the average human's lowest acuity is 200 cycles per second), and up to or perhaps even beyond the highest range of human hearing (20,000 cycles) are well within reach.

Table 5. Economics of robots.

| Feature | Humans | Robots | Cheaper |
|---------|--------|--------|---------|
| Cost | $16/hour | $4/hour | Robot |
| Reliability | 1 shift | 3 shifts | Robot |
| Safety | Bleeds | No bleeding | Robot |
| Maintenance | $120/day | $10/day | Robot |
| Productivity | 5-6 cars per | 30-40 cars per | Robot |

Touch sensors are a little more troublesome, as the touch sense is bound up with a great many other abilities, such as the ability to sense soft-hard/sturdy-fragile/moveable-immovable features of the things touched. At the MIT AI lab, there is a hand with over 200 sensors that can touch with great precision.

Taste sensors need to be able to respond to chemical compounds (fluids, solids, gases). Taste sensors can detect water and other liquids. Underground complexes at the Pentagon and the Strategic Air Command (SAC) headquarters inside Cheyenne Mountain in Colorado could well be expected to use such sensors to track and detect liquid seepage under these complexes.

In addition to sensors that can detect what humans and other animals can sense, there are many other available sensors for use in places where humans cannot both live and sense at the same time—within atomic reactors, ordinary furnaces, in handling toxic chemicals and so on. Any ordinary home has a great many sensors, some crude, some sophisticated: light switches are crude sensors because they can sense only two states, on and off; room and refrigerator thermostats sense temperature; plants sense drought; cats sense the presence of cheese and mice; water spouts sense the volume and rate of water desired; phones sense incoming calls.

The final feature of a multifunction robot is intelligence. Later on we'll consider this feature in greater detail. This is the real focus of AI work. A great many intelligent functions have been mechanized and can be done by robots, but there is still a long way to go.

## Disciplines Involved In Robotics

One way to answer the question that is the title of this chapter is to describe the sorts of people who are now working in the field of robotics. There are four general categories of workers in robotics:

15

the AI part

the electro-mechanical technology part

the computer part

the applied science and technology part.

In some cases, these parts are combined in one company, or even one individual. The distinction between these four functions is a *logical* distinction and not a physical one. A physical description, in terms of actual jobs available in the field, is presented in Chapter 7.

## Artificial Intelligence

Artificial intelligence is the area of inquiry into the principles underlying intelligence in humans and its duplicability in artificial ways. AI is the attempt to find mechanical or electronic—or even organic—machines that are capable of mimicking human intelligence capabilities.

Workers in AI are really the pioneers in robotic work. By that I mean that while robotics itself is a fusion of mental and physical (computer and electro-mechanical) technologies, the people who research the nature of intelligence on a mental level and who seek to create electro-mechanical devices that can perform intelligent functions are the AI people. They are the leaders. The results of their work filter back down to the computer and electro-mechanical technologies.[15]

This part of the robotic enterprise is concerned with "pure" robotics, with intelligence and robotics per se. People

[15]The *Handbook of Artificial Intelligence*, by Barr, Cohen and Feigenbaum, published in three volumes by William Kaufmann, 1983, is the most extensive reference work available in the field. For more direct work, see Webber and Nelson, *Readings in Artificial Intelligence*, Tioga Press, 1981, which has 31 classic papers in the field. For a layman's overview, see Paul Gloess, *Understanding Artificial Intelligence*, Alfred, 1981.

who work in this area are often found in the R&D depart-
ments of large corporations, or may even be a division of
their own, and in technically-oriented universities.[16]

There are four general areas of robotic research in R&D
organizations and in academic research organizations:

1. pattern recognition

2. problem-solving

3. information representation

4. natural language interpretation.

The research and experimentation that has been going
on for the last 15 to 20 years in these and related fields has
been marked by the presence of fundamentally different
approaches to the problems involved. Two divergent
approaches are visible in these areas of research. Each
approach is rooted in some fundamental assumption or perspective taken on the subject matter. One approach is marked
by the attempt to discover certain general features in or of
pattern recognition, problem-solving, or information representation and natural language interpretation, while the other
approach is marked by a rejection of generalities in favor of
specifics.

An example can be found in the case of problem-solving
research. In the mid-1950s, Herbert Simon, Allen Newell and
others claimed to have discovered a "thinking machine". The
first task to which this thinking machine—really a computer
program—was applied was to the problem of proving the
logical theorems involved in the logical system which formed

---

[16]The three main centers of academic research in AI are Stanford, MIT and Carnegie-
Mellon in Pittsburgh. Examples of industrial research in robotics are the NASA Jet
Propulsion Robotics Research Program, Lockheed (Aerospace), and IBM.

the foundation for information theory, electronics, and computer programming.[17]

Once Simon and Newell had been able to show that their program could prove logical theorems—a specific task—they attempted to generalize the power of the program into what they called a General Problem Solver. Part of the process of developing a program that could solve general problems was to have people write down what they did when they worked on solving problems, then to extract the general patterns involved and to embody them in a program. The general feeling was that to be able to discover the pattern of problem-solving in human beings, then the methods and techniques could be duplicated on a machine (and in a program). A general problem-solving machine could then be used as a general purpose, multitasking machine.

While the General Problem Solver approach had its successes, there never developed a general program for problem-solving in general. Some researchers, among them John McCarthy, rejected the general problem-solving approach and argued instead that the approach to developing AI and thinking machines should be in terms of specific areas of knowledge. This alternative approach has led to the development of what is today called the "intelligent assistant" concept of a thinking machine.

While Simon has developed the General Problem Solver into a program called BACON which can extract laws and regularities from input data, the approach advocated by McCarthy and others has led to the development of a variety of intelligent assistants. One version is known as Dendral which is used as a chemist's assistant to interpret mass spectographic data. Another is Macsyma which functions as a

---

[17]This logic is that of Bertrand Russell and Alfred North Whitehead as embodied in their joint book, *Principia Mathematica* (1911). This book developed "symbolic" or mathematical logic into a unified view. In 1938, Claude Shannon took this logic and showed how it could be used to formulate logic diagrams for electronic circuits. This application formed the foundation for information theory.

18

mathematics assistant to manipulate algebraic functions. This latter was developed by Joel Moses at MIT. The "intelligent assistant" route has influenced the development of medical assistants, of intelligent programs which can help medical practictioners in the diagnosis of illnesses, and so on. These machines do not embody general problem-solving techniques, but are able to apply a variety of rules— like "If . . . Then . . . "—to medical or other specific situations.

Another area of research where there has been controversy over the best approach is in pattern recognition, for example in vision research. The development of vision systems has been guided by one principal question: How are light rays impinging on the retina translated into images and concepts of things perceived? What is the relation between and how do sensations get translated into images and concepts?

One approach was to try to store images of things perceived in computer memory and then to have the vision system compare any one thing "seen" to a multitude of stored images. The problem here is that even were it possible to store an image of every object that could ever be seen, the problem of comparing an object seen to a multitude of stored images would take an inordinate amount of time. It was felt by an opposing approach that the only way in which a machine could be taught to be able to "see" and recognize things seen was to discover and develop general symbols or patterns characteristic of things that can be perceived, and to allow for variations in the patterns. In this latter approach, a small variety of general patterns can be stored in memory, with allowable variations, then an object "seen" can be compared with a small variety of patterns rather than with thousands of specific images. In the case of pattern recognition, then, the general rather than the specific approach has been found to be most useful in developing a "seeing" machine.

In the case of speech recognition, a similar state of affairs exists. Were all the words that a speaker can utter to be clear

19

and distinct in meaning with no shades of ambiguity, no context, no colloquialisms, etc., then any word uttered by any speaker could be recognized and understood by a machine without much trouble. But what if all spoken language is characterized by a great variety of tone, meaning, ambiguity, and grammatical shape?

With regard to speech recognition problems, we are currently in the "discrete utterance" stage. For example, the computer which answers the phone and responds to caller inquiries is such a machine. Basically, the machine is programmed to recognize and respond to specific sorts of terms and to a limited vocabulary. Just as in the case of vision, the approach is to store a small set of words with a variety of inflections and to have the machine compare what a caller says to that small set of terms. Even then, the comparison takes a lot of time and is very slow.

The approach that is being worked on currently is to move from discrete to general utterances, that is, to develop machines which can listen to the general variety of human speech and can recognize what is being said. Since it would be impossible to store every single word that can be spoken together with every variation in the way a single word can be pronounced or used in a sentence (including denotatively and connotatively), the current approach is to try to develop a speech recognition machine in terms of "frames" or "scripts" or general patterns.

### Bodies and Manipulators

This area of robotics is concerned with the producibility of AI creations. So, we want a robot that can jump over tall buildings in a single bound. . . . Fine, but how do we build it? What abilities does it need to have? Once we have decided that, we need to decide how to implement those abilities. Do we make the robot flexible and plastic-like so that it can

20

stretch itself over buildings, or do we give the robot great strength (like Superman)? Do we use wheels as locomotive media, or do we use stilts, or webs, or feet? Should it have wings, or hands and arms?

We can look at the space program to find examples of the more sophisticated robots, such as the one that landed on Mars (Viking Mars Lander) and performed a number of experiments and functions. In a sense, though, that robot was single-function as it never did anything else, but in principle it could have been retrieved and turned to some other task. The Mars robot was not a multiprogrammed and reprogrammable general manipulator, but it did have the capability of acting on its own in landing in an alien environment.

Another example of an intelligent robot is the Pioneer 10, the satellite which has escaped from the solar system and is now on its way to outer space. It is carrying a plaque devised in part by Carl Sagan and symbolically depicts the anatomy of human beings together with other information about the location of the planet from which the object came. This satellite is an "intelligent" robot in many respects as it has on board routines to enable it to survive its long journey into space. As we develop the intelligent capabilities of machines further, we can expect to see more such probes being sent into space to explore, map and determine locations where humans can exploit space for economic and informational benefits.

At this time, however, this side of robotics is still struggling with devices that can climb stairs, so I think that leaping tall buildings is not a capability that is just around the corner.

## Computers and Robotics

The computer revolution, and in particular the microcomputer revolution brought on by silicon chips, made robots feasible. Since chips eliminate the bulk of the mechanical mind, such minds can now be placed on devices that are fairly small. My APPLE II™ computer had in 1978 the equivalent power in a small box of an entire floor of a building that housed the earliest computers made with tubes. Making less do more is the motto of the computer industry, as prices for all computer devices have plunged. Computing power is thousands of times cheaper now than when first introduced. Buckminster Fuller used to call this the phenomenon of *ephemeralization*, of doing it better and cheaper by doing it smaller.

The process of ephemeralization is going on in three directions today. On the immediate horizon is the work being done on VLSI (Very Large-Scale Integration) chips that can contain more and more transistors on a plate of sand the size of a finger nail. The more transistors that can fit on a chip, the greater the power of the chip. Once again, the Japanese are ahead of the US in the development of fifth generation super computers. Robots are needed to make the chips for these computers, and that is one reason why Japan is ahead of us in robots.

Another area of development of chips is at the atomic level. Since electricity travels at the speed of light, computer speed is related to how far an electronic signal has to travel in order to make a computation. A chip that can handle many different computations at one time in nanoseconds is a chip that makes the distance that has to be travelled by electricity smaller and smaller.

The shortest distance that electricity would have to travel, and the place where electricity would not be slowed down (resistances) by the material through which it has to travel, is inside the atom. If chips could be built inside atoms, then computations would take place at the speed of light, and many more computations could be done "instantaneously". If we get to this level, we will in effect have chips of greater computational power than that of human beings. This can be explained as follows.

22

There are roughly $10^{11}$ neurons in a human brain, each with 100 dendrites (connections). Each connection is like a byte, a character, a knowable or useable bit of information. Thus the brain of a human being is capable of making $10^{14}$ or 1,000,000,000,000,000 "bits" of information. While this figure is astronomical, it is but a small fraction ($\frac{1}{10}$) of the number of atoms in a grain of salt. Chips based on atomic structures will not only be faster and more capable than anything we now have in silicon chips, but they will also be much smaller, so that without exaggeration we could say that an atomic computer the size of a pack of cigarettes could contain as much computing power as is contained in all computers now (1985) world-wide.

The third approach to speeding up computations is in the direction of the development of organic chips. Organic chips could be programmed with a "genetic code" that would develop very complex chips, all operating according to principles used in the brain, the most powerful computer there is.

Basically, a robot the size of a football field wouldn't be worth having; but if a robot on the human scale can be constructed, then fine. Chips made that possible.

Because of the chip, many of the capabilities required for intelligence (e.g., sensation, memory, ability to follow rules) can be housed in compartments equivalent in size or smaller than those on a human scale. Personal robots can house their "mind" and sensors in a compartment the size (roughly) of a human head. Storing a computer on or within a robot is the first step to a real robot. However, the sorts of programs needed by fully capable robots may not be the sort we presently have.

Inorganic chips are still not small and powerful enough for truly powerful robots. Organic chips, on the other hand, are living cells with built-in programs that are both much faster than "sand" (silicon) chips and also more complex and powerful. When the brain is duplicable, then robots will

23

become feasible. When the nervous system has been duplicated in the laboratory, and sensors and muscles made artificially, then we will have the technological basis for really powerful robots.

## Applied Science and Technology

Finally, the field of robotics encompasses the fields of applied science and technology. This is not so much a separate field as a separate function in the previous three areas. This area of robotics is involved in *fusing* AI with computers with mechanical technology.

Here are the features that we want in a robot:

Flexible, adaptable and programmable.

Interfaceable to computers external to themselves.

Able to learn from experience.

Able to function uniformly and reliably without external control.

Able to perform both precision (e.g., threading a needle) and also gross (e.g., throwing a stone) manipulations.

Self-powered.

Able to speak and understand speech in any language.

Able to interact with its environment, other robots, and people.

The day of the personal robot is upon us. Within the decade of the 1980s it is expected that personal computers will be common as televisions. In the 1990s, it will be the personal robot. Demagogues of the future will promise "A robot in every room." The widespread availability of robots is, however, contingent on the field of applied science and technology.

One can take AI people alone and we will not have robots. Take computers alone and we will not have robots. Take technologists alone and we will still not have them. Robots depend on the development of and integration between the three areas; applied science is the name for the function of integrating "ideals" and discoveries from AI, with capabilities and programmabilities from computing, with mechanical devices for maneuvering and manipulating. This applied science area is the function of fusing, integrating, and creating the robot, the intelligent general manipulator, that is in everyone's future.

The chart below indicates some of the capabilities that robots can be expected to have, and the time-frame for their implementation.

Table 6. Projection of robot capabilities.
Scale: * = low level    ***** = high level

| Feature | 1985 | 1990 | 1995 | 2000 |
|---------|------|------|------|------|
| Flexible | | * | ** | *** |
| Adaptable | | | * | ** |
| Programmable | * | ** | *** | **** |
| Interfaces | | * | | |
| Learn from experience | * | ** | *** | **** |
| Uniform | * | ** | **** | **** |
| Self-powered | * | * | ** | *** |
| Speech | * | * | ** | *** |
| Any language | | | | * |
| Interactive | * | * | ** | ** |
| Emotional | | | | * |
| Feeling | | | | * |
| Imaginative | | | | * |
| Problem-solving | * | * | ** | ** |
| Recognition | * | * | ** | ** |

25

## Review

What is robotics?

So far, we have suggested a number of answers to this question. A robot is:

A noun and a verb, a thing and an activity.

It is human-like but not entirely autonomous.

Robot technology is a melding of computer and electro-mechanical technologies.

A robot is reprogrammable, multifunctional and multimanipulatory.

People who work in robotics span AI, mechanical, computer, and applied science specialities.

Out of these considerations, we can derive a definition of a robot as that term will be used henceforth:

*A Robot is a not necessarily inferior mechanical subordinate of human beings.*

This means that a robot

is not necessarily inferior to humans—it may possess capabilities like human beings and some that humans don't have.

is a mechanical or artificial being neither spontaneously created nor habitual to earth.

is subordinate to humans.

is modelled on human mental and active capabilities.

# History of Robotics 2

## Introduction

Though the idea of and the attempt to create mechanical contraptions that we call robots have been in existence for possibly as long as any of the other great and general ideas, it was in 1921 that Karel Capek first published his play *R.U.R.* (Rossum's Universal Robot), a play with the term "robot" in its title. The term is a Czechoslovakian term meaning "worker". As used in the play, however, the term takes on the meaning of an invented idea.

In the 1930's, the term "android" (from *andros*, human, and *eidos*, form) was introduced by scifi writers to refer to an artificially constructed robot that, unlike a robot, had the characteristic of humans, such as being evil and unpredictable. Similarly, the term "cyborg" (cybernetic organism) was

invented around the same time and used to refer to a combination of human and machine. The term cyborg has been
replaced by "bionic" in recent times, particularly since the
TV show *The Six Million Dollar Man.*

As understood in the general scifi-high tech culture, then,
robots are distinguished from androids and bionic beings.
As established by leading scifi thinkers, there is a not necessarily acceptable distinction between bionic beings which
use artificial replacements for human organs and physical
parts, androids that are human-like and suffer from the same
fallibilities as humans do, and robots which are regarded as
somehow superior to humans. This latter view is due to Isaac
Asimov's view of robots—considered in Chapter 5—which
regards them as somehow above and beyond the realm of
humanity. This distorted view of robots needs to be considered later.

When considered generally or universally, however, it
is apparent that the term robot can be taken as the generic
or group term for all other artificially constructed *isomorphs.*[18]
I will use the term robot here as the generic or collective term
that includes single-function industrial robots, general-
purpose robots, androids, cyborgs, and bionic beings. I consider all of these as *isomorphs* of the human or mythical paradigm. That is, robots, androids and bionic beings all involve
the attempt to duplicate or improve upon or model the powers and structures of human and mythical creators. As such,
isomorphs need not have the same form or shape, nor need
they be mechanical only, or organic only. Isomorphs of the

---

[18]The term is my invention to refer to beings that have "morphs" or structures that
are "iso" or equal to those of the two paradigms of general-purpose, multimanipulatory devices that we refer to as robots. One of those paradigms is God (Allah,
Jaweh, Zeus, Buddha, Krishna, and so on) who can in fact leap tall buildings in a
single stride, and perform numerous wondrous deeds. The other paradigm of a
robot is the human being who possesses the most advanced general-purpose manipulation capabilities now in existence. My use of the term "isomorph" is meant as
a way of referring to robots, androids, cyborgs, bionic beings, and any other "creation" of any form, human or otherwise.

28

human or mythical paradigm may take any shape, play any role, be able to perform some or all or many more functions than humans.

Considered in terms of the human paradigm, a robot is a general manipulator with intelligence. Considered in terms of a mythical paradigm, robots can be modelled onto beings with superhuman powers (as Asimov seems to have done), such as Atlas or Samson or Perseus or Spiderman.

## What This "History" Involves

The history of robots and robotics, then, will be taken not as one narrow history of one little idea, but as a general idea, and thus a general history of isomorphic beings, of beings with shapes, forms and functional characteristics similar to humans and animals.

We cannot really say that preceding events foreshadowed or determined the later events. Many of the ideas related to or having an effect on robotics have been around for thousands of years. These are not ideas about robotics, but ideas that robotics, when it emerged, either took up or had an affinity for. The prehistory of robotics is not an early history, but a history of ideas that were taken up by or are related to robotics.

Real history, conscious history, emerges when people explicitly recognize and name a phenomenon or event, and then follow, trace and monitor it. Until the phenomenon of a mechanical device with reprogrammable and multifunction capabilities emerged as an idea (in the 1920s), there was no conscious history of robotics; robotics didn't exist, wasn't yet named as a phenomenon to be tracked.

Looking backwards, we can track the idea of robots, as we are using the term, in myths, and continuing in legends up to recent times. We can also track the technology of robotics way back to at least 1500 B.C. There are also ideas, or modes of thought, that influenced the later conscious devel-

29

opment of robotics, and since the term was itself invented in 1921, we can trace a development in the general culture.

We can discern any number of streams flowing eventually into the idea of robots and robotics:

- There are philosophical ideas about the nature of mind, of the universe, and of human beings.

- There are mythical and legendary ideas and concepts:
mythical beings with mythical creations;
real people with mythical creations.

- There are real people with "mythical" machines:
real people with hoaxes or single-function creations;
real people with real creations that are teleooperated;
real people with real "autonomous" mechanical creations.

- There is the technology of robots that is an attempt to actualize some of the ideas.

This history, then, will trace four roots of modern robotics. No attempt is made to be exhaustive or definitive, but only to outline pathways for study. We will take up in turn

Plato and Descartes as decisive thinkers in the history of robotics,

robots in myth and legend,

robots in technology up until 1922, and

modern robotics since 1922.


## Plato and the Algorithmic Idea

Disregarding the actual mechanical and technological attempts that may have been made to duplicate human intelligent and maneuverability capabilities, there are two main phenomena that relate to the idea of robotics. One idea comes to us from Plato. The other comes from Descartes in the 17th century, 2,100 years after Plato.

Though the idea came to Plato through Pythagoras, the idea that natural events and phenomena follow a describable pattern in their behavior is critical to the development of robotics, and of many other things as well.

Prior to any attempt to discover or map out the contours and trace the details of any situation in which one finds oneself, the whole scene may appear as a buzzing, blooming confusion. A new city, a new workplace, a new area of study, an unexplained phenomenon, all involve initial contact with disorder, dissarray, confusion. An intelligent person's first task is to try to organize and map the situation.

On the level of sensation and perception alone, things can appear confused (see Table 7). It takes an effort to search for describable order and pattern. No one is quite sure how we jump from sensation to perception, but we do know that when we perceive, something is added to sensation. Sensory deprivation and sensory bombardment experiments have shown that we operate between these two extremes; we cannot deal with total deprivation or a total inundation of sensations. To operate reliably, we have to filter out many sensations and respond to those that are "meaningful" or relevant.

For a preliterate or feudalistic society, nature is populated by many gods. In the Voodoo religion followed by many

Table 7. Development of concepts.

| Development | Perception | Conception | Stage |
|---|---|---|---|
| First | No pattern | Experience | 1–2 years |
| Second | Form pattern | Hypothesis | 2–7 years |
| Third | Test | Test | 7–16 |
| Fourth | Reject/accept | Reject/accept | 16+ |

31

Haitians, the universe is populated by thousands of gods that must be tracked, placated, fed, worshipped, etc. in order to get through life. This animistic religion based on the idea that everything is alive populates the universe with chaos, confusion and disorder. In animisms, everything has its own logic, and nothing can be explained in terms of anything else.

Plato's idea was that when you deal with the world solely in terms of sensation, everything appears to be unrelated to everything else. But when you look at events in terms of patterns and ideas, then you see connections and patterns.

This idea was applied with particular relevance to robotics in Plato's book *Meno*. The idea that he developed there is that any natural phenomenon could be understood by using a few simple logical steps:

What is this phenomenon? (Name it.)

What are the characteristics of this phenomenon? (Describe it.)

How do these characteristics relate to each other? (Explain it.)

Consider an example:

This is a triangle.

Triangles have three sides. The sum of the angles of any triangle add up to half a circle (in traditional geometry).

The relation of each side to every other side is such that in any triangle with a 90 degree angle, the square of the hypotenuse (the longest side) is the sum of the square of the length of the other two sides.[19]

The tradition of searching for patterns continued after Plato to influence mathematics in the form of, for example, Aristotle's logical works that formed the standard "text" for

---

[19]This is the Pythagorean Theorem.

logic until 1912 when Russell and Whitehead's *Principia Mathematica* started the "symbolic" logic trend in logic. This symbolic logic is the basis for all electronic theory, computer programming logic, and computer architecture.

This tradition can be called the algorithmic tradition in the sense that it constitutes a search for the "steps" to solve any problem, for the pattern that explains any phenomenon. It is a historical tradition of "going beyond" sensation, a tradition best exemplified by Einstein's theory of relativity that involved a jump beyond common sense and beyond sensory experience.[20]

Patterning must have a procedure. The history of the algorithmic idea is the idea of a procedure to solve a problem. The chief feature of this history is the emergence of more and more powerful ways of stating and of carrying out complex procedures and more and more technologies to do them. Mathematics is the best example of complex procedures that give us understanding of and control over nature.

One example should help to clarify this point. One of the most commonly used mathematical techniques today is known as the Fourier series. This technique is based on ideas from two mathematicians, Bernouilli and Fourier.

Bernouilli's contribution was what is known in probability theory and statistics as the law of large numbers, which is that given a sufficiently long period of time, and a large enough number of numeric values and measurements describing chance-ridden events, the values will fall into a pattern.

Fourier's contribution was the idea that any chance-like event or series of events could be described using the idea of a wave: the crests and troughs of an ocean wave can be

---

[20]Einstein's ideas about light, for example, are counter-intuitive, that is, they do not originate in sensory experience but in countering sense experience. The prediction that light would bend when it passes a heavy body, which was the first testable deduction of the theory of relativity, is not something that was arrived at by examining sensory experience and generalizing from it.

predicted and the heights and depths of those crests and troughs can be shown to fall into a definite pattern. Fourier's analysis of wave motions is one of science's most versatile weapons of analysis and it is hard to find a science that has not made some use of it. What Fourier saw was this. Whenever we look at the phenomena of nature, whatever they may be, we find that on first contact they appear as having no order or intelligibility. Take, for instance, the phenomenon related to trying to ascertain and predict the number of sales that can be expected to be made in a year. Looked at in terms of one perspective, the way people buy products seems to have no pattern: someone may buy one product today, another person may want many products. Looked at in individual terms, we cannot predict what any one individual will buy. But when we look at the whole sales phenomenon in general, we can see that there are patterns: for all sales that can be expected to be made this year, we can see that for every product there is a level at which sales seem to be made, and a trend-line that shows whether or not sales are going up or down. Looked at as an aggregate, the sales phenomenon makes sense and can be patterned. This is a key idea of the scientific community.

The two decisive events in the algorithmic tradition were (1) Pascal and Babbage who showed that "mind" could be mechanized, and (2) the electrification of algorithms for mental machines (computers). This was the end of a long tradition that had begun with Pascal's attempt to mechanize thought.

Robots combine the electrification of algorithms on mental machines with procedures developed on physical machines.

## Descartes and the Mechanical Idea

The second idea of importance in the prehistory of robotics is the idea that comes to us from Descartes.

Whereas Plato claimed, in effect, that the place to find

34

order and pattern was not in the natural, physical world, per se, but in a mental world, Descartes in effect said that the natural world is patterned also, and that on looking at anything physical, we can describe the mechanical principles that operate in it. For Descartes, the mind has patterns and the body has patterns that can be described.

In his 1662 book *De Homine*, Descartes describes humans and animals as machines, that is, as mechanical and artificially creatable devices. Descartes advanced a theoretical model of mechanical man that certainly had an influence on subsequent work in this area.

If a pattern can be described, then it can be duplicated and reproduced. Not only may we duplicate and reproduce the patterns of the mind, as Plato held, but we may also duplicate the patterns of the body. Bodies can be understood in purely mechanical terms. This being so, we can create machines that can duplicate those bodies. A hundred years later, another Frenchman, LaMettrie, wrote a book describing humans as machines (*L'Homme Machine*).

If the body, the mind and (some of) its operations can be mechanized, then we in effect have here in the 17th century the idea at the basis of robotics. The conscious history of robotics, which did not start until the 20th century, is in effect but a series of footnotes to Descartes (1596–1650). Descartes in effect took mind and regularized its operations, took the body and mechanicalized it, and ever since then, the question has been: How are we to describe and duplicate one or the other in mechanical form?

## Robots in Myth and Legend

There is a factual and there is a mythical or legendary development of mechanical beings. The factual development we will consider in the next section. The mythical or legendary development concerns us here. By mythical or legendary, we mean simply that there is no evidence that such beings

35

existed in fact, and that in the absence of such evidence, the best that we can say, at least of the alleged creations of human beings, is that they foreshadowed in ideas and creations that human beings were later to attempt and in some cases succeed.

The best sources for factual details on the history of robotics can be derived from works on the history of technology. Several of these are listed in the bibliography.[21]

## Robots Created By Divinities

The myths and legends of Greece are filled with references to the divinities who created isomorphs of various kinds. The Biblical story of creation may be regarded as the story of a divinity who created the "next best" robot. What we will consider in this section is the creation of mythical robots by mythical beings.

In the Genesis story of creation, Yahweh formed a human out of clay and then breathed life into it. That is, Yahweh created a physical reality (out of clay) and got the design to act as a living thing (breathing). All the divinities have this power of "life" and "breath". Similarly, all myths and legends relating to humans who (allegedly) created isomorphs supposedly had the power of "breathing life into" an otherwise inanimate being. Anyone who could literally bring someone back from the dead was at one time regarded as at least divinely related, while today modern medicine routinely brings people back to life. Since this is so routine, it is those people who are able to bring people back from a different "level of reality" who are regarded as divinities.

---

[21]A good place to begin and a source used throughout the remainder of this chapter is the book, *Robots, Robots, Robots,* edited by Harry M. Geduld and Ronald Gottesman that contains a remarkable variety of robot-related articles, including ones by Carl Sagan, Mary Shelley, Arthur Clarke, and many others.

In Greek legend, a clear instance of divine creation of an artificial being is found in the Pygmalion story. That story has three main acts: (1) Pygmalion falls in love with Aphrodite; (2) when she rejects his love, he constructs an ivory statue of Aphrodite and makes love to it instead; (3) Aphrodite is so touched that she "breathes life" into the statue and thereafter, Pygmalion has Galatea, a living statue, as his love. (Genetic engineering can result in *organic robotics* where the result will probably be a profusion of living robots, organic isomorphs capable of growing, learning, changing.)

Aphrodite's husband, Hephaestus, created female servants made of gold and able to speak. Daedalus, a descendant of Hephaestus, was reputed to have made statues that could move of their own accord.

These are all clearly examples of mythical beings who created equally mythical or imaginative beings which clearly represent *deep* wishes in the human psyche. We have here (see Table 8):

statues coming to life

speaking creations

servant girls

movable creations.

Table 8.

| Mythical Beings | Mythical Creations |
|---|---|
| Yahweh | Humans |
| Aphrodite | Enlivens a statue |
| Daedalus | Servant |
| Hephaestus | Gold servants |

## Robots Created By Humans

In this section, a different emphasis prevails: we will consider real human beings who created mythical beings in the sense that there is no evidence that these "real" people did in fact create workable robots. Table 9 summarizes the chronology of these alleged creations.

Empedocles, a philosopher living in the 5th century B.C., is the first human being to be credited with having made an isomorph. He is alleged to have animated a statue.

Another example comes to us many years later, sometime between A.D. 1021 and 1058, when a Spanish nobleman is said to have created a female isomorph to do his housework. A few years later, Albertus Magnus, a priest, was said to have spent over 20 years constructing a robot made of wood, metal, wax and leather that was fully mobile and could welcome visitors at his door and speak to them. According to legend, the fellow who is now Saint Thomas Aquinas is said to have destroyed Magnus' robot on the grounds it was the work of the devil. This was sometime between 1193 and 1280. (This is just another sad instance of the clash between inventors and Luddites. St. Thomas seems on this score to be something of a Luddite.)

Roger Bacon, 1214–1294, is said to have created a head

Table 9. Chronology of mythical robots.

| Chronology | Real Humans | Mythical Creations |
|---|---|---|
| 5th C. B.C. | Empedocles | Animated statue |
| 12th C. | Albertus Magnus | Servant girl |
| 13th C. | Bacon | Talking head |
| 16th C. | Loew | Golem |
| 16th C. | Paracelsus | Little man |
| 17th C. | Goethe | Robot |
| 19th C. | Anderson | Mechanical birds |

that could speak. Paracelsus, 1493–1541, is said to have constructed "little men," homunculi, as a result of his discovery of the philosopher's stone, a secret akin to the power of "breathing life" that belongs only to the gods. Another reputed Cabbalist and occultist, the Rabbi Loew (1525–1609), supposedly used a magical formula to bring a clay robot to life. This animated clay giant is reputed to have protected the Jews against a foreseen pogrom. This animated clay robot and isomorphs to it are referred to as the golem.

Robots have played a role in literature. Goethe's *The Sorcerer's Apprentice* deals with robots in parts. The well-known stories of Sinbad the Sailor in the *Arabian Nights* stories are populated with automatons of various sorts. One robot kills two gravediggers; other automatons resemble singing peacocks. Hans Christian Andersen populated some of his stories with mechanical birds.

Of course, Mary Shelley's *Frankenstein* of 1818 has had a vast influence on the more "philosophical" side of robotics where people spend much energy discussing the merits and demerits of any attempt to create artificial beings; the pro-technologists see these creations as beneficial, whereas the Luddites see them as terrifying in their potential for evil and destruction.

This difference is a long-standing one, rooted in the very foundation of Western society (see Table 10 for some examples). The Hellenic and Roman attitudes were that machines are useful and beneficial artifacts. The Hebraic tradition, on

Table 10.

| Support Technology | Luddities |
|---|---|
| Albertus Magnus | St. Thomas |
| Marvin Minsky | Hubert Dreyfus |
| High Tech | Many labor unions |
| People who hope | People who fear |

Table 11. Benefits of technology.

| Economic Benefits | Old Technology | New Technology |
|---|---|---|
| Workers | Many People | Fewer people |
| Products | Fewer Products | More products |

the other hand, regards machines as wicked and heretical.

The Luddite versus technological advancement conflict is a perennial one. The term "Luddite" is a 19th century term, but the idea is ancient. Beginning in 1811, and continuing for five years thereafter, a group of craftsmen who called themselves Luddites took it upon themselves to try to destroy industrial civilization. The principal reason they had was that they feared that industry would destroy their jobs. In the 16th century, no one would buy Pascal's calculating machine. The reasons advanced were of two kinds, one legitimate, the other not. The legitimate reason against Pascal's machine was simply that since Pascal was the only one who knew how to keep the machines working, then no one else could maintain the machines. The illegitimate reason given was that the machine would eliminate jobs.

Every technological advance in human history has been opposed on grounds such as these. The clear economic rule is that new technologies create more jobs, but this is ignored by the Luddites (see Table 11).

**Summary**

In myth, legend, and farfetched but possible worlds, the following capabilities have been attributed to artificial beings:

life and animation

speech

servant capabilities

movability.

40

# Mechanical Robotics Up to 1922

The history of mechanical robots that really existed is a fascinating one. In this section, we will again only touch on a few highlights of this history.

Here we thus consider real humans who created real robots. This is the third leg of the history of robotics: real people with real machines.

## Mechanical Robots BC

Around 1500 B.C., Egyptian water clocks supposedly used human figurines to strike the hour bells. This is the case of a mechanical (as distinct from electronic) algorithm, a procedure to get the figure to move in a regular way and to perform another procedure, e.g., strike a clock. Lots of ingenuity was involved: water, air, hydraulics were used to do many "physical" actions.

In 400 B.C., Archytus of Tarentum, who is reputed to have invented the pulley and the screw, two indispensable tools, is also said to have invented a wooden pigeon that could fly.

The second century B.C. in Hellenic Egypt was a time of the development of many automata. All over the place there were statues (puppets?) which were said to be able to speak, gesture and prophecy. Hero of Alexandria invented a hydraulically operated statue of Hercules slaying the dragon in the third century B.C. Hero was an early Buckminster Fuller who wrote on mechanics, invented the slot machine and steam engine, and who constructed an elaborate theater using water to move the actors into and out of their scenes.

In the second century B.C., Philo is said to have made an even more elaborate theater that could go through five whole acts of a play from beginning to end.

The principal characteristic of all these mechanical creations is that they are single-function, single-program creations.

41

A Layman's Introduction to Robotics

## Mechanical Robots Since A.D. 1

The development of mechanical robots since A.D. 1 has been phenomenal. The machines that have been invented to duplicate human capabilities or as isomorphs for them have been varied and ingenious.

In the first century, Petronius Arbiter created a doll that could move like a human being. Considering that children and adults were faddishly interested in talking dolls only a few years ago, it is remarkable how old the idea really is.

For example, Ismael Al-Jazari, a 12th century Arabic engineer, wrote *The Science of Ingenious Machines* providing the details and designs of machines that had existed in imagination or fact for thousands of years. A few years later, we find Leonardo da Vinci, after years of trying to develop a wing that could, when added to the human, make him fly like a bird, giving up on the idea of a flying machine and turning in 1510 to the development of a mechanical human, supposedly on the basis of the idea that mechanical humans are easier to construct than bird-like humans.

In 1557, Giovanni Torriani made a wooden robot for an Emperor that could fetch his daily bread from the store. Sometime between then and 1650 when Descartes died, he and Christian Huygens are said to have constructed robots of various kinds that demonstrated the "mechanical" aspect of human beings and all animals, and thus their duplicability.

The eighteenth century was the heyday of automata before this century. At every turn, complicated and ingenious—though for all practical purposes useless—robots were constructed.

Vaucanson (1709–1782), for example, created in 1738 a mechanical duck that could eat, excrete passable iso-olfactoric excrement, walk, quack, and do various other duck-like things except fly. Later on, Vaucanson constructed a flute player that could play many different pieces of music.

Another example of a robot, this time single-purpose, is

the young writer created by Droz (1721–1790) that could write a one-page letter and then sign its name at the end. This machine was constructed using a technology similar to the talking-book recordings of the early days of recorded music where mechanical pegs on a round "player" were struck in sequence on a sound-emitter. It would not be very difficult today to re-fashion the mechanical "player" into an electronic machine capable of being programmed. A human being could then "say" the words, and a little fellow could do the actual "handwriting". This little fellow has been replaced by the word processor or text editor that is one of the most common pieces of office-related software on office computers.

The nineteenth century was also privileged with robotic inventions of many kinds. In 1890, Edison developed yet another version of the talking doll; in 1893, the Canadians developed an elaborate steam-driven robot with an exhaust extruding from its mouth; and in 1897, the French film *Clown and Automaton* was filmed.

But it is to 1875 that we must turn to see what could have been the most elaborate mechanism of that century. J.N. Maskelyne's *Psycho* made its debut in London in that year. This was a mechanical "halfman" sitting at an elaborate "desk" who could nod his head and perform mathematical feats of addition, subtraction, and division (using principles already developed by Pascal in the 17th century). Psycho could also do some low-level conjuring tricks, could spell, smoke cigarettes, and could play Whist*.

The thing about Psycho is not that he (she?) could play bridge, but that he was able to *win* when playing against humans, not just once or twice, but win thousands of games against just a few dozen losses over a span of years. Unfortunately, this device was later unmasked as a hoax achieved by running pneumatic tubes between the mechanical man

---

*A card game, similar to bridge, that involves probabilities and strategic skills.

43

Table 12. Real humans and real robots.

| Time | Creator | Robot |
|---|---|---|
| 1st C. | Petronius Arbiter | Moving Doll |
| 16th C. | Leonardo | Mechanical man |
| 16th C. | Giovanni Torriani | Walking Robot |
| 17th C. | Christian Huygens | Various Robots |
| 18th C. | Vaucanson | Flute Player |
| | | Mechanical Duck |
| 18th C. | Droz | Writer Robot |
| 19th C. | Edison | Talking Doll |
| 19th C. | Canadians | Steam-driven robot |
| 19th C. | French | *Clown and* |
| | | *Automatons* |
| 19th C. | Maskelyne | Psycho |

and a human hidden from view. Even so, ingenuity and mechanical inventiveness was involved here. Today, instead of pneumatic tubes to control the robot, we use electronic phone lines and computers. The difference is that we don't pretend that the robot is really doing "all that" itself. Machines controlled from the outside are called "teleoperators" and perform many useful tasks. Psycho was a teleoperated machine, a very clever one. We do not today condemn people for using teleoperated machines, we praise them.

These examples (see Table 12), all indicate the continuing interest in robots, yet the lack of the development of mechanically capable robots. Clever and ingenious people developed clever and ingenious machines that, though remarkable, were one and all but poor shadows of the powers that machines have today. Just as 99 percent of all scientific discoveries have been made in the 20th century, so 99 percent of all robotic developments have been made in the 20th century.

# History of Robotics Since 1922

1922 is just an arbitrary date. Since the invention of the term robot by Capek in 1921, the conscious history of robotics has grown in leaps and bounds. This development has two sides, a popular one and a theoretical one. The popular side of robotic development since 1921 has to do mainly with the "fantasy" industries—film and scifi—while the theoretical side has mainly to do with developments in electronics and mechanics, and in the logic of robotics.

# Robots in Science Fiction

Up until *The Wizard of Oz* of 1939, films about robots seemed mainly to be about mechanical servants. In 1909, the British produced a film called *The Electric Servant*; in 1924, another film was addressed to *A Machine that Thinks*; while in 1937, another film focussed on the *Mechanical Handy Man*.

*The Wizard of Oz* with its Tin Man who lost and tried to find his heart seemed to change the focus of robotics from the slavish side to its more "equivalent" side. For example, the TV series *Star Trek* created robots that were as or more intelligent than humans in some cases, and certainly a lot more powerful. Similarly, the movie, *2001: A Space Odyssey* set several hundred years earlier than the setting for *Star Trek*, depicted in 1968 robots of great power and capability.

The 1973 movie *The Six Million Dollar Man* that became a TV series developed the theme of humans aided and abetted by mechanical means to develop and increase their powers. Mangled beyond recognition in a burning plane, the hero was put back together using technologies that gave him superhuman powers.

# Modern Theoretical Robotics

The history of robotics is tied up with that of computers, engineering, electro-mechanical technology, and other areas.

45

A Layman's Introduction to Robotics

It is difficult to say exactly where one ends and the other begins. In this section, only a rough chronology of significant events that seem to be involved in robotics will be presented.

Three topics will be dealt with:

1) the background development of the search for a computing machine

2) significant events in the development of modern computers

3) developments in AI.

**Prehistory of Computers**

An early version of an information processing machine is the Chinese abacus, invented in 2600 B.C. but in existence since 5000 B.C., that uses many of the same principles employed in electronic data processing. In the West, the earliest interesting piece of information technology was Blaise Pascal's machine first built when he was 19 as a way of getting out of the tedium of helping his father with his calculations. This was in 1642. All he wanted was something that anyone today can afford—a hand calculator. He thought way back then that a calculator could help his father add, multiply and subtract figures in a less time-consuming way.[22] Pascal's machine, of which more than ten are known to still exist, was the first advance in computing devices in 4,000 years. (Since the abacus, there had been other attempts made to construct cal-

---

[22]Interestingly, up until 1940, three hundred years later, the idea seemed to be that a calculator, while useful to the father of a genius son, was not generally useful.

An amusing event that occurred prior to 1940 was when a leading inventor-developer in the growing computer revolution went to his managers at the Bell laboratories and said something to the effect, "Gentlemen, for $50,000 I think I can build you a general-purpose relay calculator."

The response that he got from the management at Bell Laboratories went something like, "Gee, who wants to spend $50,000 just to be able to calculate."

culating or computing machines, the earliest being Ramon Lull's Ars Magna, an early logic machine.)

The next advance in calculating machines was built by Leibniz in 1673. An admirer of Pascal's machine, Leibniz used his own creative logical and mechanical ideas to construct the first general-purpose calculator.

The next advance in developing a machine that could perform calculations was the analytical engine of Charles Babbage (1791–1871) in the 1800s. Using logical symbols and ideas developed by one of his contemporaries, Auguste DeMorgan, and leaning, no doubt, on his lovely companion Lady Lovelace, Byron's only legitimate daughter, he spent his life trying to get a general calculating machine to work. His chief difficulty lay in the fact that metal gears and parts cannot do the processing job adequately. Babbage needed the development of electronics to realize his dream.[23] Babbage's machine could do all the arithmetical calculations, and could do 60 additions a minute, but was made of metal and wood, and was never able to work properly. Like modern computers, Babbage's machine had four sections: (1) a place to store data (memory banks, disk, tapes, internal memory in modern computers), (2) a mill (which is what we today call the processor), (3) a place or activity that transferred data between storage and processor, and (4) input and output capabilities.

In the early 1930s, Vanevar Bush tried to develop a calculator based on analog principles. His differential analyzer (1942) was not accepted into the technological mainstream, but was nonetheless yet another attempt to mechanize the calculating processes, an attempt begun several hundred years before by Leonardo and Pascal.

The next advance in the development of information processing technology began to take shape in the late 1800s

---

[23]See Philip and Emily Morrison, Eds., *Charles Babbage and His Calculating Engines*, NY: Dover, 1961, Pb.

and early 1900s. What occurred during this time was a great surge in mathematical formalism, and the feeling that everything could be quantified and calculated, a feeling shared, of course, only by a few mathematicians.

This process of increasingly symbolizing and formalizing mathematics resulted in the publication in 1910–13 of *Principia Mathematica* by Bertrand Russell and Alfred North Whitehead. Using both logical as well as mathematical insights, they developed the logic, "the formal way of thinking and solving problems," that has become the foundation for most of the work and thought that goes on throughout the culture generally, and specifically the logic of the information industries.

The logic of *principia mathematica*—technically called predicate or second order predicate calculus—is based throughout on two key operations: *And*—used to join bits of information together, and *Not*—used to deny that one bit of information belongs with another bit.

In the electronic theories that began to develop in the 1930s and 1940s the central developments in switching theory—which formed the basis for the transition from tubes to transistors, and later still to chips—were developments that used as key operators *And-gates* and *Not* or *Or-gates*.

The developments in electronics were related to developments in computing that were taking place due to the mathematician, John Von Neumann and the logician Alan Turing, both of whom were the geniuses behind the development of working electronically-based calculating machines. Von Neumann's mathematics tied in nicely with the logic and mathematics needed by the electronic theorists. Turing made his greatest contribution to computing in the form of a theory of "computability".

A related contribution to information technology was made by Claude Shannon (interestingly enough, in 1948, at the Bell labs), who directly influenced two different areas of the industry. On the one hand, he influenced the way computers are constructed, the way the electronics works, the way the machine interprets its signals and commands, by selecting the *bit* as the unit of information. On the other hand, he influenced the whole development of communications (and telecommunications) technologies in the "Mathematical Theory of Communication" that he developed and published in 1948.[24]

## Modern Computer Developments

Some milestones in the history of information machines are:

1927—The first successful electronic computer (analog) of Bush.

1938—Shannon's papers on using symbolic logic on electronic circuits.

1939—IBM starts work on a computer.

1944—Howard Aiken's Mark I computer.

1946—ENIAC, the first digital computer using tubes instead of electronic relays (as in Bush's computer).

1948—Science of cybernetics is born.

1949—Shannon develops information theory.

1950—EDVAC computer implements the "stored program" concept.

1951—UNIVAC, the first commercially available stored program computer.

1952—IBM's 701 computer is marketed.

---

[24]Shannon's papers.

1954—IBM markets models 704 and 705 with 4k internal (RAM) memory. No self-respecting microcomputer today is without 32k or 64k in memory.

1956—IBM invents Fortran (a programming language) to help sell its 701.

1958—Using transistors instead of tubes, IBM's 7090 and 7070 computers for commercial data processing are introduced.

1964—Using chip instead of transistor technology, IBM introduces the IBM model 360 computer, the machine that gave IBM its status as the giant of computing machines.

1975/76—the first microcomputers were being put together.

Now, we are in the midst of an ever-expanding information revolution.

## AI and Computers

In the previous two subsections, we saw the development of computers and, subsequently, the development of computer science and computing machines of greater and greater power. In this section, we will examine the development of AI and robot-related developments. While computers represent the machines which can be used to accept, store and execute symbolic manipulations, AI represents the area which explores the problem of the types of symbolisms that can be formulated, in particular the symbols involved in intelligent behavior.

An early significant event in the history of robotics was Alan Turing's computability theorem of 1936. In a very technical but fairly easy-to-understand article, Turing explained what it meant for something to be mechanically "computa-

50

ble". This was also a significant event in the history of computing.

The second significant event occurred with "The General and Logical Theory of Automata" by John Von Neumann, the mathematical genius who had a great deal to do with the development of computers. This theory, presented in a recent book (see bibliography), but in development since the 1940s, is based on the idea that a *real* robot is one that can reproduce itself. A real robot, in the full sense of the term, is an artificial being who is able to create another robot identical to itself. Once a real robot is created, we can take it into the middle of the desert and command it to duplicate itself. The process leads to the multiplication of robots, each new one itself created with the instructions to create one like itself for as long as the programmer desires. A great many significant ideas have been based on the ideas of Von Neumann.

In the mid-1950s, Herbert Simon and other early pioneers in computers felt that something better should be found to do with computers than simply using them to crunch numbers (still their primary function). So, in 1956, an AI laboratory was opened at Dartmouth College with Marvin Minsky.[25]

In 1957, the miniaturization of electronic components began with the invention of the transistor. I can remember my thrill from being able to stop dealing with bulky and persnickety tubes to transistors in my electronic projects in high school. This process of miniaturization, or *ephemeralization* as Buckminster Fuller calls it, is significant not only in electronics but also in the mechanical parts and tools that we construct today. Tools are precise to a much greater degree than before; tolerances are measured in microns rather than millimeters; the speed of computers is measured in nano-

---

[25]See Betsy Staples, "Computer Intelligence: Unlimited and Untapped," *Creative Computing*, August, 1983, Vol. 9, No. 8, pp. 164–166.

seconds; thousands of transistors are etched into each silicon chip.

The next significant event was the institutionalization of AI by the development of the AI Lab at MIT in 1965 by Minsky, and the opening of a similar lab at Stanford around the same time. Both labs have the same aim—to develop concepts and devices with vision, mobility, hand manipulation, and intelligence. In other words, the aim is to develop an "independent agent" that can function in the "real" world, and not simply in the abstract world in which computers function.

In the early 1960s silicon chips became feasible, thus further reducing the size of electronic components and increasing the amount of computing power that could be housed in a small space. In 1975, the first microcomputer was marketed.

In 1979, a significant event took place with the appearance of the magazine *Robotics Age* (Summer, 1979). The 66 pages of Volume 1, No. 1, dealt with topics like digital speed control, industrial robots, robot vision, chess-playing robots, and robotics in the Soviet Union. One of its sections was devoted to a section on competition among readers to come up with "useful" things that robots could do or could be expected to do in the future.

In the 1980s, the Japanese showed the world what personal and single-purpose robots could do to improve productivity in manufacturing concerns.

## Technological Robots

The practical application of robotics is today being done mainly by industry. While academic work focuses on theoretical and experimental areas, industry has taken the discoveries made and the machines developed and has tried to turn them to practical advantages.

## Industrial Robots

The first machine for use in industry was a manipulator, a hand device with an end-effector produced by Seiko in 1939. This machine did not achieve a high visibility or use (it is still around). That was something done by Unimate in 1958 that was sold to GM. By 1981, GM had 35 such robots and Chrysler had 24.

Unimate was invented by George Devol who holds the patent on the machine. The honor of having promoted this machine into the public consciousness belongs to Joseph Engelberger, President of Unimation, the largest producer of industrial robots, who was influenced by Asimov and who is regarded as the father of industrial robotics.

Today, Unimation's PUMA robot is regarded as the most advanced industrial robot on the market.

## Personal Robots

The history of personal robotics is even shorter. In 1980, the Terrapin Turtle appeared. This was a small robot with the capability of vision, talking, movement, etc. It was created by a young student who had studied at MIT.

The first major development in the field was the appearance of the HERO-1 personal robots from Heath, followed shortly by RB Robot Corporation RB5X robot. Soon came B.O.B./XA from Androbot Corporation.

The First International Congress of Personal Robots was held in Albuquerque in April, 1984, and the First International Conference of Industrial Robots was held in June, 1984. The decade of the robot has begun.

# Robotic Concepts 3

## Introduction

Imagine yourself in the familiar environment of a large department store with a 360-degree view. As you look around you can see things that you recognize—clothes, cameras, computers, plants, and so on. The *concept* of a department store from your point of view are the items that you see as relevant to your needs.

The concepts of the department store when seen from the point of view of the manager, or owner, may be quite different. For the owner, the principal concepts may be income (caused by customers), outgo (caused by inventory), and the difference between the two resulting in loss or profit. For the manager, the basic concepts may be stock on hand and in transit, customer demand, and the like.

These differences suggest that concepts and the things that they refer to are *relative* to an interest or point of view. This consideration of the relativity of concepts is a static, three-dimensional one. If we now add the element of time, and consider the matter under four dimensions, what we see is a river of flowing concepts, some sinking, some coming to the surface. For the consumer, the central concepts are different from those of the owner or manager. What is central (on the surface) for one group may be below the surface (of no relevance) to the other.

Concepts are areas of sustainable interest: features, traits, characteristics of events, environments or objects. What concepts are described for robots depends on the point of view taken on them. Theoreticians will have one view; practitioners another; laymen yet another. No one set of concepts exhausts the topic; all are legitimate. The concepts of robotics would also be described differently by the mechanical or electrical engineer, by the user of robots, or by the businessperson or vendor.

In this chapter, we will attempt to describe the concepts of robotics in terms of a framework—a point of view—related to the capabilities that robots may have. That is, instead of defining the concepts of robotics in terms of, say, mechanical or engineering concepts which often have to do with the components of robot mechanisms, we will deal with the concepts in terms of the *end product* and what the end product is expected to be able to do.

The purpose of this chapter is thus to attempt to sum up the central concepts which define the "boundaries" of robotics. Most of these concepts have been hinted at already in the previous two chapters, though some additional ideas have to be brought in here. Though it would be quite feasible to divide these concepts into groups, I have chosen here to consider the concepts in terms of their role in relation to the various functions that robots are expected to perform.

The concept of a robot, as we have seen, has two dimensions: the dimension of algorithms for mental (intelligent) procedures, and the dimension of physical "algorithms" for executing those procedures. To describe the concepts of robotics in terms of the capabilities of robots is to describe the concepts in terms of the intelligent (mental) and physical (auto-motive) capabilities.

## Overview of Robot Capabilities

To do the actions a robot is ideally expected to do, robots need:

A body with end-effectors so as to be able to perform.

Sensing capabilities so as to be able to gain knowledge of itself and its environment.

Perceptive capabilities so that it can recognize and identify, and above all differentiate, the phenomena it encounters.

Memory capabilities so it can learn from experience and recall successful solutions to previously encountered problems.

Rule intelligence so that it can be expected to act in a consistent manner.

Creative intelligence so that it need not be told everything it must do.

"Emotions" so that it can understand and sense the emotions of living beings.

Feelings so that it can appreciate the significance of events.

"Imagination" so that it can develop foresight.

Intuition so that it may sense the environment in different ways.

The level of the "body" is the level at which all pre-electronic robots were constructed. The aim throughout the history of mechanical and mythical robots that we looked at

57

in the previous chapter is the history of attempts to construct mechanisms similar to natural organisms using mechanical means to duplicate the external and apparent way that things work.

The development of mechanical means to sensation really began when electronics came on the scene starting with radio in the early years of this century. Sensation depends on the development of sensors, on mechanical means to sense that something or other is the case in the environment. This capability is pretty much hardware-oriented. One sort of sensor is litmus paper that senses the presence of acidity in a solution. Cats' whiskers are also sensors.

When it comes to perception, things start to get a bit harder. The perceptual level is where AI begins. It is at this level that purely hardware or mechanical means cease to be adequate, and some software or intelligent procedure must be found in order to mechanize perception. This area of robotic capabilities is still in its infancy.

The area of robotic memory is pretty far along as the computer industry has developed memory capabilities of great power. This is pretty much a hardware development area depending on computer technologies.

The areas of rule and creative intelligence are being investigated on the frontiers of AI research. This is where AI efforts are concentrated in an attempt to mechanize "reasoning".

The four capabilities of emotion, feeling, imagination and intuition, all right-brain capabilities, have not yet been subjected to as intense study as the previous areas. It is not at all clear that robots need these capabilities; even more to the point, no reliable means has been found to mechanize these capabilities. In all likelihood, the first six capabilities will be developed over the next 15 years, while the last four will take a back seat. After the turn of the century, however, we can expect to see the development of feeling and imaginative robots.

You may say that there is in fact no need for robots to be able to feel, and that perhaps the whole set of capabilities called emotion, feeling, imagination and intuition do not need to be developed for robots. However, consider the following: In order to be able to "care" for something, the carer must be able to have some feelings towards the caree. An emotionally neutral being (e.g., a robot) could not be expected to be able to understand the value I place on my child, the care and attention I expect it to receive, and why, unless it has some feeling capabilities. This is one reason why feeling is included and will be considered in this chapter.

Each of these ten capabilities will be discussed in turn.

## Robot Bodies

To be able to *do* the sorts of things that robots are supposed to be able to do, they must have bodies. Human bodies are self-contained enclosures based on a jointed skeleton with two arms, two legs, and various internal organs to act as life-support systems. If they are to be multifunctional, and if they are to perform as complements to human beings, then robots will have to be able to move effectively the way humans or animals move. They need not be two-legged (biped) nor two-handed as we are, nor need they have the same life-support internal systems: We need stomachs, they need storage batteries.

In addition to physical means of locomotion on land, robots as "complements" to human beings should be able to perform feats of locomotion that combine the appropriate features of other machines. Some robots may be expected to be able to move as easily on land as through the air, sea, and even outer space. At the present time, we're lucky if a robot can map out a room and pick up small objects, but this can be expected to change as robots take over more and more of the activities that humans have performed up to now.

An early image of a robot is HAL in the movie *2001*.

59

Other robots and androids have appeared in the popular *Star Trek* of TV and *Star Wars*. Some imagined or even constructed robots have been given human "features"; others have the capabilities without the features; others have their own separate capabilities.

Robots with human features are today manufactured by one of the personal computer manufacturers, Androbot Inc. Their B.O.B. robot has a humanoid body, for example. Running on a very stable complex of three wheels, B.O.B. has three megabytes of memory, an arm, plus optional peripherals like speech, security alarms, telephone answering services and others.

Heath's HERO-1 robot, on the other hand, looks like R2D2 of *Star Wars*, and though lacking a humanoid body it still is constructed to perform human-like functions. It comes with speech capabilities and an arm. The RB5X model is nonhumanoid, but comes with the same capabilities—an arm, speech, and other capabilities.

It doesn't much matter, really, what the body of a robot resembles, so long as that body can do what it is supposed to do. If someone houses a robot in the body of a bank vault, he should not expect it to be able to work in a rose garden. I fully expect that after "natural" bodies and their capabilities have been used on robots, other "forms" for bodies will be invented that will have the capabilities wanted, but that will look entirely unlike any existing natural body. For example, the body that may be suited to all environments, all localities, all types of conditions (water, fire, hail, etc.), and capable of many different tasks, may be one that has many different arms and types of locomotive devices, that can collapse itself or expand as the case may be, that can separate itself into two parts momentarily (e.g., to hold onto the "other" side of an object), and so on.

So far, the construction and nature of bodies with locomotive and functional capabilities is hundreds of years old in the development of "bodies" that can be housed in large places, but "little" bodies haven't been developed very far. We can build space vehicles and projects with ease, but we cannot duplicate the functions of any ordinary animal. The development of androids, of machines that resemble and duplicate the actions of animals, will lead to revolutionary changes. Imagine being able to have an android cat with all the abilities of a cat, yet at the same time possess the mental capabilities of a human. Imagine the possibilities this creates for all the human vices and crimes, along with espionage and other such activities. . . . One can simply send the "cat" out to do one's dirty deeds.

It is not impossible to imagine that at some time in the future human beings will have surrounded themselves with intelligent plants and androids for all species on earth. Being able to be "on the same" intellectual level as an eagle, earthworm, or whale would give humans capabilities for making discoveries about this planet unheard of before.

So we have humanoid bodies, android bodies, nonanimalistic bodies and extra-animalistic bodies. Nonanimalistic bodies include plants and extra-animalistic bodies include bodies with the shape or function of all nonorganic "bodies" on earth. It is, again, not unimaginable that one day we could have intelligent snowflakes, intelligent rocks, intelligent houses.[26]

## Robot Sensations

In addition to just having a body, robots need to have sensory capabilities. These capabilities separate a "dead" animal body from a "live" one. Unlike computers which respond symbolically, robots are expected to be able to sense their environment, gather information about it, and respond not just

---

[26]The evolution of what could well be called an intelligent house can be traced in many articles by Steve Giarcia in BYTE magazine over the last five or six years.

to abstract electronic signals, but to respond *to* the environment and to change their behavior in light of that information.

Sensors allow interaction with the environment in the form of fact gathering, orientation, the evaluation of problems and the development of solutions.

In schematic form, the simplest picture or model of a robot in terms of its sensory capacities can be seen in Figure 2. An environment is responded to and sensed by available sensors which are themselves interpreted by the central processing unit, the computer.

As we have already noted, sensors of many varieties have already been built though their capabilities are minimal. There are aural sensors that can detect sound with a reso-

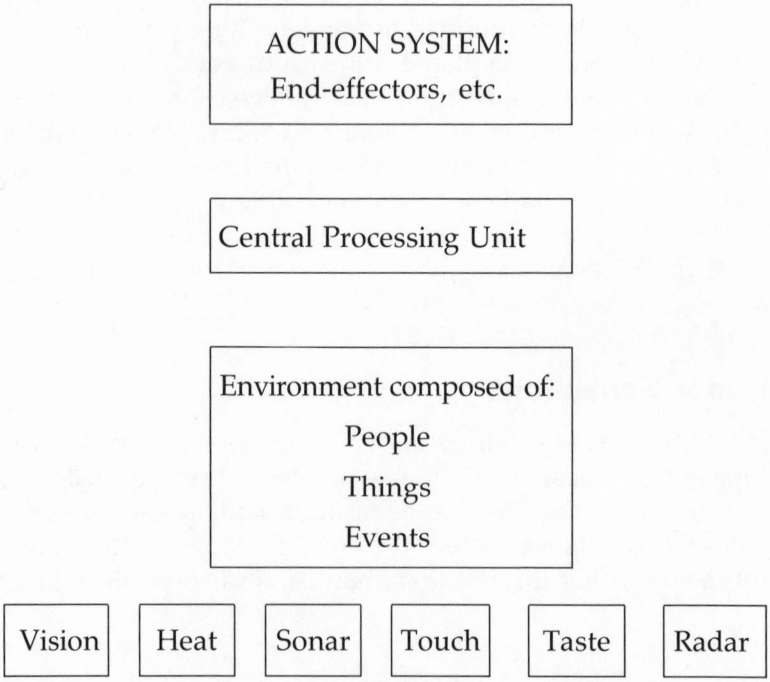

```
┌─────────────────────────────┐
│      ACTION SYSTEM:          │
│      End-effectors, etc.     │
└─────────────────────────────┘


┌─────────────────────────────┐
│   Central Processing Unit    │
└─────────────────────────────┘


┌─────────────────────────────┐
│   Environment composed of:   │
│          People              │
│          Things              │
│          Events              │
└─────────────────────────────┘
```

| Vision | Heat | Sonar | Touch | Taste | Radar |

Figure 2. Functional model of a robot.

lution of 1 in 256, 1 in 512, maybe 1 in 1024. Similarly, existing light sensors for seeing can detect ambient light in the visible spectrum with the resolution equivalent to that of a TV screen. (However, the military has vision and sound detectors with much greater power than this.) The TV cameras and screens used by satellites is very high resolution so that much clearer and detailed visions are possible, but military systems are very expensive, whereas robots are marketable only if their price is right. This is in the range of $50,000 to $100,000 for industrial robots, and $5,000 to $10,000 for personal robots.

There are sniffing robots which can detect the presence of water and other liquids, and there are robots with thermovisual scanning devices, tactile senses, ultrasonic senses and so on.

Touching sensors have been in development, but still need a lot of work to be able to detect differences—soft/hard, etc. Gripper mechanisms that can hold objects are thus not yet highly developed, though work is advancing rapidly. There are robots that can build electronic instruments, bathe hospital patients, feed paraplegics, grip atomic wastes. The Space Shuttle arm is very strong even if not extensively maneuverable, though it proved its usefulness in snatching a broken-down satellite, bringing it into the Shuttle's cargo bay for repair, then placing it back into orbit (April, 1984).

In addition to the normal five senses, there are sensors that can detect sensations of varieties that humans are not capable of, e.g., at higher or lower levels of hearing, x-rays, and so on. The light spectrum, for example, extends all the way from gamma rays (high frequency waves of small wavelength) to x-rays, ultraviolet waves, visible light, infrared, and radio waves. Our eyes have evolved to be able to sense the visible part of the light spectrum (the seven colors of the rainbow: red, orange, yellow, green, blue, indigo, violet). The reason why our eyes see only in the visible spectrum may be that most of the sun's light is focussed on that region. For robots, however, we can make them sensitive to the whole range of the light spectrum to which we are not sensitive. With gamma ray sensors, robots could detect radiation hazards; with x-ray sensitivity, they could, like Superman, be able to "see through" solid objects; with infrared sensors, they could see at night, and with radar and radio senses they could detect signals and objects normally unsensed. It is only a matter of evolutionary accident that we were not all born with x-ray vision as was Superman.

## Robot Perception

It is one thing to be able to have sensations and detect sensory phenomena. It is quite another to perceive these phenomena. To sense is not to perceive. Eyes take in the environment and the brain synthesizes and organizes it. For example, 600 trillion light rays strike the eye every second, but we certainly do not "sense" 600 trillion separate light rays. We sense, instead, groups of rays.

To perceive is to be able to detect the pattern in the sensations. It is known that human beings have brains that "filter" out most of the sensations that we experience. Human beings experience sensations—sensory stimuli—at the rate of 100,000 per second. Only a few of these are left over after the filtering process, so that in most cases, even the most aware human being is simply not "normally" aware of the greater part of his sensory experiences. Sensory overwhelming and sensory deprivation cause breakdowns in orientation and the ability to behave normally.

The filtering process is what we are interested in. What we "perceive" are sensations as interpreted (subliminally) and as meaningful. We perceive patterns of sensations; pain, for example, is not a single sensation but an ongoing series of sensations of several types. Perception thus involves both sensation and interpretation, and that in turn requires representation, symbolization. That "clump" of sensations over there is a table; this clump over here is a chair, for example.

64

In AI parlance, the whole field of perceptual research is termed "pattern recognition" research. Robots and computers have been given the capability of sensing but not yet the capability of perception, except to a low degree. The HERO-1, for example, can detect still and moving objects, but cannot without additional clues identify the nature of these objects. Being able to describe the patterns involved in perception and being able to teach robots how to "perceive" those patterns (how to filter, in other words) is a major area of research in AI.

There are three principal types of vision systems. One type is the binary vision system. A second is the ternary vision system. A third is the structural or topological vision system.

The binary system uses just two "values," light and dark. This system assigns to each pattern (e.g., a cup) a position in a grid-like system. Each coordinate is assigned some value, such as light (1) or dark (0). Each position on the grid is then stored digitally in the computer. Whenever an object is encountered, the computer has then to compare the object encountered with the model or picture of objects stored in memory. This is a clumsy way to approach the problem as this method requires that every possible object that can be encountered in the environment must have a stored picture in a computer memory. This method yields an infinite number of pictures that must be stored, and the time taken by a computer to compare any encountered object with the stored pictures would be considerable.

A second type of vision system is the ternary system, so-called because it includes a wide range of values from light to dark, that is, it includes *shades* between the light and dark values. Instead of two values, each dot can be one of 256 or 512 (or more) shades. The value of this vision system is that it allows the development of more complex templates or models for objects and allows vision systems to recognize more complex objects.

65

# A Layman's Introduction to Robotics

The third type of system is the structural or topological type. This alternative approach is now being developed. It is based on the idea that there are structural similarities between all objects that can be encountered, and that the primary form of recognition is recognition of these structural similarities. Once the structure of an object is recognized, then variations in the structure are compared with the object in the environment.

Pattern recognition studies encompass a broad range of individual efforts. One main area of work is in the development of speech synthesizers that can "utter" human speech. This technology is making good headway, and all articulated human speech can be duplicated by a speech synthesizer mechanism. The other side of this coin is the ability to sense human speech and to interpret and "understand" the language correctly. So far, though great strides have been taken, there is no program that can help a computer or robot interpret spoken language correctly most of the time. The range of inflection, pronunciation, tone, and so on is so great that at this point, only "straight" and limited talk can be understood. For example, the computerized voices that answer the phones at banks, telephone exchanges, and so on can understand and respond to only a limited vocabulary. Robots and computers can easily speak, but they can't easily recognize and interpret speech they "hear," or they can't do it fast enough.

For example, the mean for speech translation programs is 100 minutes to "translate" 30 seconds of spoken speech into understood speech. This is how long it takes to "hear" and "interpret" talk in general. The robots which can answer the telephone interpret by using key words and then using a preprogrammed sequence to respond. They do not really "understand" or interpret talk.

There is a difference between hearing an indistinct sound and hearing a sound indistinctly. An indistinct sound, a barely audible word spoken in haste, an unfamiliar sound echoing

66

in a valley—these are sounds that we forgive people for not perceiving, and we can forgive robots for the same thing. But perceiving a sound indistinctly shows an incapacity—a hearing loss or mental incapacity to perceive at all.

As this is being written, there are industrial robots which can perceive one simple part, like a nut or bolt, when it is laid out against a background. But if the part is mixed in with a variety of other parts, then nothing can be "seen". The latest vision systems can detect when two parts are touching each other. This involves "seeing" both parts and detecting touching or nontouching.

The same sort of problem exists with language. This problem with language pattern recognition extends to all the other senses as well. To be able to "see" as distinct from merely sensing light, a robot needs to be able to identify objects in its environment. Given 100,000 sensory stimuli a second, let us suppose, how do you teach a robot to filter out the sensations that "belong" to objects like chairs and tables, and to identify the patterns of these objects? In short, robots can detect *that* something is there, but not necessarily *what* it is.

To detect *that* something is there requires only a simple contrast mechanism. To identify *what* something is (to perceive it) involves images and concepts and words. It is one thing to *sense that* something is there, but quite another to *perceive what* is there.

Beyond this is the need to recognize the *importance* of what is there and of how to deal with it: One does not need to watch a refrigerator, but one does need to watch a boiling pot. We cannot tell a robot (or computer) everything that it is to watch for. Rather, we need to give it rules to help it figure out for itself what needs to be watched and what doesn't, and what the problems are that it may encounter in its environment.

An example is the problem of recognition. Robots today can recognize one part on a definite background, but being able to "see" one part among many is difficult to do and has not been achieved. What we need here is to know the type of knowledge involved in perception and how to program a robot to perceive. While a robot today can deal with a "stable" environment where there are simple objects to be perceived against a stable background, we have not yet been able to program a robot to be able to deal with the different scales of objects, of objects that rotate or that appear different as they are rotated, and thus the difference in perspectives. A robot may recognize a plate when looking at it "straight on," but when looking at it sideways where the perspective is different, this is not yet possible.

## Remembrances Of Robots Past

Robot memories are as advanced, obviously, as computer memories. Memories allow a robot or a computer to have access to and be able to store, use, and recall a great many different things at once. Memories are measured in the industry in terms of increments of 1024 bytes, referred to as 1K bytes.

Robot memories range from as small as a few K, 16 or so, for some of the personal robots, to three megabytes for Androbot's robots. Having to store information about procedures, maneuvers and decisions all in one place requires memory storage. The technology is constantly developing ways of storing more bytes for less money and in less space. Chips, disks, tape, bubble, laser are all technologies that have been put to use. If, presumably, there is a reason why humans have the size brain we do, even though we apparently don't use very much of that power, then perhaps there is a relation between the amount of things one needs to store in memory in order to be able to do the things that humans do, and the size of the memory capacities.

There is an illness called aphasia that results when parts of the brain literally rot away. People lose their memories (and thus their abilities to speak) and their abilities to perform elementary functions. Take away selectively, e.g., through implanted electrodes, parts of the memory in humans, and we lose the abilities to recognize, identify, hear, see, speak, walk, and even to breathe. How much we really need, and how much correspondingly a robot needs, is something that we do not yet know. Here it is not so much a question of how much memory, as the memory of what and how is all the memory to be related to its parts.

A really general-purpose robot is probably going to need at least 3000 megabytes of memory to be able to talk, understand, see, hear, perceive, and act intelligently.[27]

The problem here is this. To be able to duplicate any of the human powers of perception and conception so as to interpret and understand events in the environment, a robot— a computer-on-wheels—will need two sorts of storage mechanisms. One is a directly accessible memory called RAM for "real access memory" which also needs some external storage mechanism to store numerous files of data and programs that need to be called and run as applicable. In addition to these two common elements of computer storage, the most important thing needed is the ability to access and retrieve information quickly, almost instantaneously.

And here is the rub. According to Einstein's theory of relativity that governs all physics, including the physics of computers, the one absolute in the physical universe is that no event can exceed the velocity of light. The electronic mechanisms used to store and access information in computer memories are subject to this limitation on the speed of light. The whole revolution in computer technology brought about by silicon chips and more recently by the VLSI chips are based on attempts to circumvent this limitation. If electrons

---

[27]The figure of 3000 megabytes is arrived at as follows. There are $10^{14}$ neuron-dendrites in the human brain. This is 1 million billion. It is estimated that humans use only 3 billionths of their brain power. This is equivalent to 3000 megabytes.

travel at the speed of light, then in order to be able to make more computations quickly, we need to find a way whereby the electron processing mechanisms do not require those electrons to travel very far. The more transistors that can be put on a chip and the smaller the chip, the more computations that can be done at one time, and the less distance the electrons have to travel to accomplish these computations.

If we were somehow able to create chips where the distances between transistors were reduced infinitesimally, and if we could bypass the use of metal and other conductive substances that impede the flow of electrons, then we could have a chip that could handle infinitely many computations at the almost instantaneous speeds of that of light. This is the chief reason why investigations are now proceeding to create atomic-level chips. If we could create atoms that were computers, then we could have computers going on in a "small" area much, much smaller than silicon chips, and much faster as the electrons would not be slowed down by having to travel through metal conductors.

Until we get to the level of atomic chips, then we are going to have to deal with the limitations of existing technology. This means that space is needed for memory storage, and that the number of computations that can be done at any one time will be limited by the chips available.

What that 3000 megabytes or more consists of is something else altogether. Do we store all the words in the *Oxford English Dictionary*, not to mention dictionaries for all other languages? Do we store the procedures for every single function such a robot could ever perform (which would be a waste of time), or do we develop general procedures that can be applied by an (intelligent) robot to each specific situation as it perceives them to be at the time? And if we do store general procedures, we need to know what these procedures are, and how do we construct machines that can understand general procedures and adapt them to any situation? How do we teach a robot to recognize subtleties of meaning in words, or to understand colloquialisms? How can we avoid sending our personal robot on a wild goose chase when we tell it, "Go to hell"?

## Rule Intelligence

The fifth capability that a robot needs is rule intelligence. Rule intelligence is, as the name implies, the intelligence to follow rules. The computer, for example, is a rule machine without peer. Computer programming languages are symbols and words used to "tell" a computer what to do. In most cases, every step, and every twist of every step, must be specified, for the computer does exactly what it is programmed to do, and nothing else.

This is one reason why computers have been programmed to play chess so well. The rules of chess are simple— Do unto others before they do unto you. Given that goal, the strategy is to explore everything he can do to you, and everything that you can do to him, in every situation, and choose the most productive move.

The way humans play chess, so it has been found, is to consider at most 20 or 30 different possible moves in any one play situation, and then to select one of those, even though after 10 moves in chess there are many thousands of possible moves at each juncture. How do humans manage to eliminate many thousands of moves at one fell swoop? One way that I have not seen talked about as such is that when facing a chessboard, humans instinctively divide the board up into time-phases and sectors or categories of moves, so that when they look at the board, they always "see" at least seven categories of moves at any one time, and they know that the other thousands upon thousands of moves fit somewhere into those categories. For each category, one considers the most promising moves.

A computer, on the other hand, doesn't time-phase or sectorize moves, but instead considers at each step every

71

possible move it could make, and every possible consequence, before it selects a move. Present day computers are sequential processors. A significant new development is parallel processors. A parallel processing computer can perform a multitude of computations at one time and thus is closer to the actual functioning of the human brain. Because it can work so much faster on the detail level than human beings can, it can process all those millions of moves in a few minutes or seconds.

Computers do it by following the rules; humans do it by breaking them. The logical rule is, when faced by a problem, define the alternatives. If there are two alternatives, define them; if there are 100, define them too; and if there are millions, do the same. Of late, chess-playing computer programs have been developed which can sectorize in a way similar to the way humans sectorize.

Medical diagnosis programs and some experimental research programs, most notably one called Bacon (developed at Carnegie-Mellon University), have been developed whereby the computer can be fed a lot of unrelated facts and can find patterns in those facts. One astounding example must suffice. All the data that Kepler collected over 20 years or more, and which he himself searched in order to discover his laws of planetary movement, was fed into Bacon's databanks, and Kepler's laws were discovered in a few minutes.[28]

Bacon is an induction machine. It can take "facts" and "sensations" and can search them for patterns, regularities, repeatabilities. Induction is the process of going from particulars to generalities, from facts to laws, from experiences to theories. To get a machine to do induction, you need facts, and you need to tell it what sorts of things to look out for, what sorts of patterns are relevant. If doing cryptoanalysis, you tell the program to search for patterns of repeated letters and so on. If doing "meaning" analysis, the number of letters

[28]See Betsy Staples, in *Creative Computing*, p. 166.

aren't relevant. If doing logical analysis, the letters don't matter, the words don't matter, just the structure of the relationship between the statements matter. In any case, once one has given Bacon an idea of what to look for, it can chug along looking for patterns forever.

Human beings are often characterized as rule-intelligence beings. Some people have created dictatorships and totalitarian systems in an effort to make human beings act like rule-followers. Some people see the key to all education as the development of habits (physical rules) of various sorts. Other people see good citizens as those who can follow rules. Much of our moral sentiment is based on the idea that a good father, brother, mother, parent, child, is one who is reliable, uniform, predictable, and of good moral conduct.

Rule intelligence is, however, a low form of intelligence. The ability and willingness to follow rules is a feature of ants, but is it really one for humans?

## Creative Intelligence

Creative intelligence involves knowledge of the rules and the ability to *deliberately break them*. There are people who know the "rules" instinctively and "break" them without conscious awareness of what they are doing. Cats left unfed too long and children being naughty both do this. Occasionally in daily life we come upon utterly charming people who are rule-breakers without knowing it, or knowing what it means. They are creative in a sort of protean sense only.

Other people don't know the rules, and don't break them anyway. This is where the great mass of the population happens to be. Other people break the rules, but don't know what the rules are, so they are not being creative, while yet other people both know and break the rules. (Other writers on creativity notwithstanding, that is what creativity means to me and will be taken to mean here.)

Let's consider the issue of creative intelligence in terms

73

of the four levels of logical operations that psychologists like Piaget[29] have discovered among people and in the stages of growing up.

There is, first, the sensori-motor stage. In this stage of logical development (the stage in which a child resides up until 2 years or so), the main logical or relational or structural capability is the ability to organize space into regions (which is akin to dividing a chessboard into sectors), and reversibility. For a robot or a computer, this level of logical ability was surpassed with the very first robot, the ENIAC in 1944.

The second level of logical power is the pre-operational (2 to 7 years) stage and level where, though there are no high level invariants and structures, there are symbols and ways of representing them. Computers can represent and store numbers and characters (symbols) and can relate them in various ways, e.g., sorting them into alphabetical order.

The third level and stage of logical operations is that of concrete operations. This level adds the powers of recognizing and dealing with classes and relations of operations and data. At the present time, relational databases that implement relations of operations and data are only just beginning to appear in the marketplace. The ability to handle classes is related to the mathematical theory of groups and to the AI problem of pattern recognition, neither of which has found a satisfactory mechanical representation.

While the third level in Piaget's schema is related to what I called rule intelligence, the fourth level is related to what I have called creative intelligence. The fourth level is a level of structural operations. It involves abilities to take data and information of any kind and to perform structural operations on that data, such as:

---

[29]Piaget's theories about mental development are outlined in just about all of his books, but particularly those dealing with the development of thought in children.

74

hypothesize about it

relate the data schematically

carry out permutations of the data

create combinations

aggregate the data in different ways

correlate the data, etc.

Creative intelligence involves the third-level/rule-intelligence ability to understand that pears and oranges don't taste the same, or that oil and water don't mix, and to ignore the rules anyway.

Idleness, reverie, play, and so on are activities that violate rules of various sorts, and yet which are vital to the continuation of the species. Idleness allows the mind to wander (and wonder), yet it violates the rule against unemployment. Reverie takes the conscious mind beyond the clear light of its narrow confines and into the pre-, sub-, un-, non-, or supra-consciousness, and thereby violates the rules against "irrational" thinking. Play allows spontaneity and invention, both of which violate the rules against emotionalism and originality.

When robots reach this level, then they will be truly intelligent. Until then, even a machine can follow rules . . .

## Emotion

Though in ordinary language we commonly use feeling and emotion interchangeably, the two are not the same. Emotion refers to a capacity to be affected, to have or sense something. Feeling, on the other hand, involves an active capacity to make a judgement, an intellectual determination, of a feeling.

Can robots have emotions? Can machines have emotions?

Somehow, people aren't too upset knowing that computers and robots can calculate and compute. But the idea that they can or will also have emotions seems to be a threat to some people. The reason is that somehow or other emotion is seen as a human response to the environment, not something that any old machine can do. Given this prejudice against emotional machines, it is a common observation that "robots can't have emotions" and that robots will never be able to have emotional experiences.

Nonetheless, the research frontiers of robotic inquiry give every indication that the problem of mechanical emotions and of emotional robots have been addressed and that work is proceeding to develop robots with emotional capabilities. In some quarters it is expected that by the turn of the century emotional robots will be available.

What this means is that robots will be able to be affected, to have passions, to be sensitive, to have "gut" reactions, to be subject to impressions that "move" them. It means that at some time, we can expect robots that are excitable, that can be agitated, that can be expected to show eruptive responses, that can be warm, touchy, and even irritable.

But this projection also has its problems. Almost without exception, in all of science fiction robots are all right until they acquire emotional capabilities; then they turn against their human masters. Once robots acquire emotions—so most of science fiction seems to suggest—that is when they set out to destroy human beings. In his 1950 book, *I, Robot*, Isaac Asimov described a "race" of robots devoid of emotion that were programmed in such a way that built into the very structure of their "being" was a set of prohibitions and laws (see Chapter 5) that would require robots to act consistently and without passion. Passions, emotions, are such frightening phenomena for some people that the aim is to eliminate or control these passions entirely. Asimov's aim seemed to be to eliminate the possibility of emotion from one of the creations of human beings.

76

# Feeling

Similar expectations are available with respect to feelings. Robots which can feel may also be part of our future.

To feel is to discriminate, to make distinctions, to be selective, to have a mental attitude, a point of view, a frame of reference in terms of which emotions are encountered and judged.

We may be emotionally affected by encounters with various people or events, but we develop feelings when these affections are warmed over by reflection and conscious development. Though erotic love is an emotion, for example, conjugal love is a developed feeling. The fear that is experienced in the face of danger is an emotion; the courage or cowardice shown in the face of that danger is a feeling.

# Imagination

Imaginative robots are something else altogether, however. The power of imagination in humans refers to the capacity to turn "inward". It involves a capacity for creative thought, for the ability to have thoughts, opinions, concepts, or mental impressions. It refers to the capacity to have "in mind" likenesses, images, resemblances, copies or models of things. It also refers to the capacity to presume, to conjecture, to infer, and to postulate, i.e., to go beyond the "actual" to the "possible".

All of these capabilities presuppose the capacity to be conscious, to be aware and to be aware that one is aware. It may even involve the capacity to have a "self," to have an "ego". At this point, we are touching on the very borderline of inquiry. In science fiction, we already have images of "conscious" robots from 2001 and from other movies and books. The question at this point is whether we understand human conscious capacities and if we can mechanize them.

## Intuition

Though talked about extensively in popular literature, intuitive abilities have only recently been subjected to scientific inquiry. As in everything else, we have two sides here: those who believe that the intuitive abilities that people claim to have should be investigated, and Luddites who say that since intuitive abilities don't exist anyway, only fraudulent inquiries can be made.

The intuitive abilities refer to experiences of foreknowledge and clairvoyance to so-called second sight or the sixth sense, and to the ability to have premonitions, hunches, and so on. Intuitive abilities also refer to the possibility of telekinesis, psychokinesis, teleportation, levitation, telepathy and so on.

There is a question as to whether or not humans actually have these abilities, and if they do, there is the additional question as to whether or not these can be mechanized and duplicated by machines such as computers and robots. At this stage, no one really knows the answer, but I suspect that like everything else, the best approach is to subject the matter to investigation rather than dismiss it outright.

## Review

Human beings are multijointed bipeds with a variety of powers and internal support systems. Externally, we have locomotive devices (legs), manipulators (hands), communication devices (hearing, sight, tongues), intelligent devices (brains), detection devices (sensors), and emotional and intuitive powers.

As a complement to human powers, robots can be expected to be mechanical counterparts to humans with mechanical analogs to the human powers. That is, if human beings, and the personhood of human beings, is regarded in terms of the powers and capabilities noted throughout this

78

chapter and in the summary above, then it is not impossible that at some time in the not-too-distant future robots will mimic these powers.

Were robots able to do so, would they then become human-like persons subject to all the rights and responsibilities of human beings? Some people say yes; others say no.

Those who say yes see no reason why robots should not at some time in the future be able to qualify as persons. Those who say no claim that robots will never be able to qualify as humans. For these latter people, to qualify as a person, robots must be able to move, reproduce, perceive, feel, learn, understand, interpret, analyze, discover, choose, and create. Now that we are on the verge of developing "expert systems," i.e., artificial intelligence capable of doing many of these things, it is questionable that robots will never be able to do these things. This is why those who say no would go on and add, well, maybe robots can do all of these things, but they cannot qualify as humans, because only humans can lie, laugh, cry and commit suicide. In this case, those who say yes simply say, if we ever create or develop a robot that can do these things, would they then qualify as humans?

The answer seems to be that no matter how capable robots may become, those who say no, those who refuse to accept robots as persons, will always find yet one more quality of human beings that robots cannot supposedly mimic, while those who say yes will always be able to find or develop robots that are capable of doing those very things. This suggests that the controversy is not one that is carried out on empirical grounds, that is, is not really a matter of whether or not robots can or cannot do something, but lies on a different, more philosophical plane. There is an *a priori* species chauvinism abroad to the effect that humans are the only persons, that nowhere in the rest of the universe are there persons, and that no matter what intelligent beings we create or discover, even if this discovery were made of beings in

79

distant space, these beings would not qualify as persons. On the other hand, those people who think that the universe has possibly billions of stars with perhaps millions of civilizations (or planets with living creatures) are more willing to accept that robots may someday be capable of being persons.

Only time, not argument, will tell.

# Robotic Hardware **4**

## Introduction

The hardware of robotics refers to the machines, the units, the mechanisms that are used.

We can generally distinguish between the mechanisms themselves as end-items (as actual consumer products) and the components or parts that make up these end-products. An industrial robot and a personal robot that are both assembled and ready-to-go are end-products. DC magnetic motors, transistor drives, positioning systems, sensor products, solenoids, grippers, accomodators, tactile end-effectors, feedback transducers and the like are components or parts that may be included in an end-product.

Here we are not concerned with the components of robotic machines such as resistors, chips, solenoids, motor mechanisms and the like. Rather, we will be considering the actual merchandise, the wares, the durable items of commerce available in the field of robotics.

Four main topics are dealt with here:

First, a general overview of robot hardware needs is provided. This overview can serve as a "standard" or "desideratum" for robotic hardware.

Second, we consider the mechanics of industrial robot mechanisms.

Third, some examples of industrial hardware products are described.

Fourth, the hardware available for personal robots, hardware available at prices affordable to individuals, is described.

## Robot Hardware Needs

Given the outline of robot capabilities presented in the previous chapter, we can say that a useful, general-purpose, multifunction, intelligent robot should have the following general capabilities:

- Sensory capabilities
      should be able to hear with great acuity;
      should be able to feel and touch;
      should be able to taste;
      should be able to smell;
      should be able to see.
  All of these capabilities should be greater than or equal to human capabilities—e.g., night vision, animal-level hearing, sensitivity to various impressions humans are not normally sensitive to (carbon monoxide, poisonous gases, and so on).

- In addition to these sensory capabilities, a worthwhile robot is going to need "perceptual" capabilities—the ability to identify and discriminate objects. This is tied in to the ability to pattern, and also the ability to interpret.

- In addition to sensory and perceptual capabilities, a robot needs to be able to *execute* commands, directions, etc. It must be maneuverable and able to manipulate

82

and handle materials of all sizes and shapes. It thus needs a "body" that includes—

hand, arm and wrist maneuverability;
memory so as to remember what to do;
coordination and agility;
reliability, speed,and executability.

- As if these capabilities were not enough, a robot also needs to be able to communicate in terms of the principal mode of human communication—speech.

- It needs to be able to think and create.

- Finally, it may need to have emotional, imaginative and intuitive capabilities.


Given this idealized image, someone could ask if they are possible, and even if they are possible, why are they needed? Why do we need robots that can do these things? Do we really need such elaborate mechanisms? The answer is yes. We need such mechanisms for many reasons, two of which are perhaps central: the exploration of outer space, and for home and industry.

We need such machines for space exploration for a simple reason. While teleoperated machines could do work in near space, where humans on earth could direct their activities, once we get much beyond the moon we run into trouble. The reason is that it takes 2.4 seconds for a command signal to get from robot to earth and back to robot on the moon. For highly complex and dangerous activities, this may be too long a time to wait. If a robot is about to step off a cliff, 2.4 seconds may be too long to wait to know what to do next.

The problem is aggravated when the distances are longer. Communication between a robot on Mars and a human operator takes 40 minutes per signal. In order to carry out exploration on the planets, and certainly beyond the solar system,

83

we need intelligent machines that can decide on their own what they need to do to solve problems encountered in their environment.

With respect to home and industrial robots, we need very capable machines that can eventually replace human labor. Chapter 8 describes the range of actual and possible uses for personal robots.

## Industrial Robots

Of the many ways in which the types of robotic hardware could be categorized, the distinction between industrial and personal robots is used here. Industrial robots are robots used in industrial—mostly manufacturing—situations, while personal robots are intended for use by individuals in their homes.

Industrial robotics[30] is in an ever-increasing development and application. Ever since the Japanese pioneered the widespread use of single- and multi-function robots on the factory floor (which happened only a few years ago), and showed the increased productivity and efficiency that could be obtained from using robots,[31] American industry has been trying to catch up. In 1980, there were at most 1,500 industrial robots at use in the US, as compared to 80,000 in Japan. Other industrial robots are in use in Volvo plants, and in France, Germany and Israel.

American robotics is not strong in the Japanese sense of wide use, but is strong in the discoveries related to the technology of electro-mechanical robots. However, there is movement afoot that may lead to explosive growth in the industry

[30]See J. Michael Callahan, "The State of Industrial Robotics," in BYTE, Vol. 7, No. 10, October, 1982, p. 128 ff. See also Joseph Engelberger, Robotics in Practice: Management and Application of Industrial Robots, Amacome, 1980. Engelberger is the president of Unimation, the leading manufacturer of industrial robots in the world.
[31]See the book by Servan-Schreiber in the bibliography. See also Robert W. Hall, Zero Inventories, APICS Publication, 1983.

in the immediate future. There are currently 100 or more US companies in robotics, with a number of computer-based companies gearing up for a plunge into robotics. IBM, for example, has a long-term robotics research program. One of their most recent products is the delivery of AML (A Manufacturing Language), developed in 1975, for use on the IBM PC. The leading producer of industrial robots is Unimation Inc. which has over 5,000 robots at use around the world, more than all the robots that exist and are in use in this country. This company expects a good share of the 50,000 or more robots that are expected to be in use in this country by 1990.

Though there are manufacturers that have produced personal robots capable of being mass produced (see next sec-

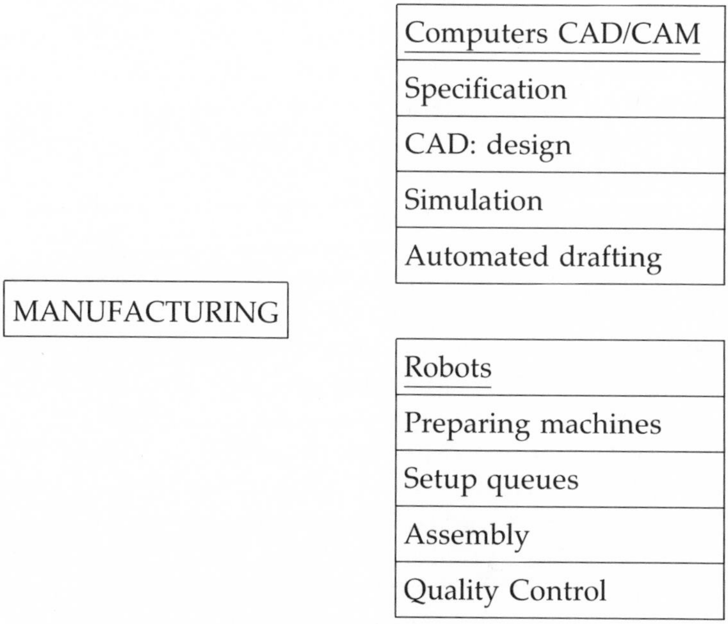

| Computers CAD/CAM |
| --- |
| Specification |
| CAD: design |
| Simulation |
| Automated drafting |

MANUFACTURING

| Robots |
| --- |
| Preparing machines |
| Setup queues |
| Assembly |
| Quality Control |

Figure 3. Robots in manufacturing.

tion), the most popular use of robots today is in industry, particularly manufacturing. Robots are used to weld, cast dies, load machine tools, spray paint (e.g., cars), for heavy lifting, for dangerous tasks, and in other ways as well.

The manufacturing industry represents the second major chunk of the US economy after the service industries. The manufacturing industry has made extensive use of computers in the form of transaction systems (e.g., order entry), decision support systems (for production and inventory control and CAM), and computer-aided design (CAD). Now that robots are on the scene, we can expect to see a similar widespread use of robots (see Figure 3).

## Components Of Robot Mechanisms

Only limited use has been made, and development attempted, of industrial robots that can use sensors to respond to an environment. In general, industrial robots, while they may be reprogrammable and multitasking units, are programmed to perform a specific task at any one time, and are given different sorts of maneuverability depending on the primary task at hand.

Sensor technology is rather limited and undistinguished. Video cameras, proximity sensors, and tactile sensors are used in helping a robot to position itself to perform a task and to help in the actual performance of those tasks. Infrared sensors are used to sense the location (or presence) of objects to be placed on pallets, for example.

Similarly, the development of perceptual capabilities and of pattern recognition capabilities is still in its infancy. This is an area in which AI work is being done. The problem to be solved is to develop (1) sensors with great acuity, and (2) to develop software that can be used to filter and pattern sensations. There are sensors available that can help a robot detect the difference between standing still and being in motion and that can recognize objects belonging in either of these two classes. However, face recognition and so on is beyond the power of any modern robot. At this stage of the game, getting a robot-computer to "see" one part set on an undifferentiated background is a cause for ecstacy. If you want multipart recognition and sortability, forget it for now.

A robot is maneuverable and versatile to the extent to which it has an adequate *power/drive system*, and to the extent to which it has the ability to *do* something. The objects that enable robots to "do" things are called, technically, *end-effectors*. Robots needs to be versatile in what they do, and highly maneuverable to be able to position themselves to do these things.

The power/drive mechanisms of robots use air (pneumatics), liquid (hydraulics), or electricity (motors). It is not unimagineable that at some time in the future, other forces may be used. The choice between these three bases for the power/drive units depends on the load and the accuracy with which movements are to be effected.

Pneumatic drive units are not used when accuracy and precision is required as air is difficult to control. However, pneumatic drive units are cheap, light in weight, and though of little strength, they are quick and cost in the range of $5,000 to $20,000.

Electric motors are used where accurate movements are required and where small power units and light loads are expected. Electric drive units are most expensive, in the $80,000 to $100,000 range.

Hydraulic methods are used wherever heavy loads and gross accuracy is needed. Normally one would expect that a truly versatile robot would have all three methods available for use at any one time.

**Manipulators**

In addition to the power/drive part of a robot, there must be a tool that is able to do things. In most if not all cases, this tool is one that is isomorphic to a human arm and wrist. The power/drive unit positions the robot to do something, and the arm-wrist does it.

The most common end-effector is a grip mechanism called a manipulator. Technically, a manipulator has a multidegree of freedom, an open-loop chain or series of mechanical joints and linkages that are driven by actuators and which is capable of *moving* an object from point to point along prescribed trajectories. Let us clarify the terms in this definition.

Imagine yourself floating in gravity-free space. Consider your nose as your center of gravity. Now, there are certain "directions" toward which you can move. Imagine a straight line extending above and down away from you vertically. Call that line the z-line. Extend your arms straight away from your body in left and right directions. Call an imaginary line along your extended arms the x-line. Now imagine a straight line stretching forward and backwards; call this the y-line.

When space, or the arena in which movement takes place, is considered in terms of what mechanical engineers would call the Cartesian (after Rene Descartes who discovered the system) coordinate system, position and motion can be expressed in terms of these x, y and z lines. Using their intersection as a point of reference, we can illustrate this coordination system as in Figure 4:

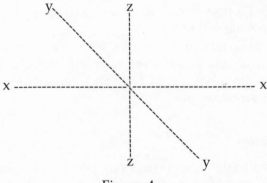

Figure 4.

A degree of freedom is motion along one axis between the point of reference and one of the six directions: left, right, forward, backward, up and down. A really useful robot will be able to move along each axis or coordinate and thus will have six degrees of freedom. Many industrial robots have five or fewer degrees of freedom, however.

A conveyor belt has one degree of freedom—it can only move forward in one direction (see Figure 5).

Conveyor Belt

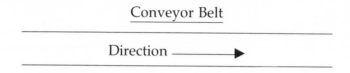

Figure 5.  One degree of freedom.

A bicycle can go backward and forward; it has two degrees of freedom (see Figure 6).

Figure 6.  Two degrees of freedom.

Of course, there are also three, four, five, and six degrees of freedom movements.

An actuator is a motor that causes motion in response to a signal. A manipulator has joints (like finger joints) and links that are joined together in a series of mutually responsive and interacting motions.

Movement is simply a change of position or rotation around a point. When a movement takes place, we need three bits of information to understand and explain it. We need a *device* which is to move relative to a point of reference; we need a goal or *working point*; and we need to specify the *kind* of motion that is to take place.

89

For example, the device might be a gripper; the point is to move from point A to point B. The kind of motion (trajectory) might be one of several types: (1) movement in a straight line, (2) circular motion, or (3) rotation.

A manipulator is thus a device that is able to rotate or change position (relative to a point of reference) using one of three trajectories, moving in some combination of the degrees of freedom in order to go (move) from point A to point B, both points specified in terms of the coordinates x, y and z.

When we deal with the software of robotics in the next chapter, the Cartesian coordinates will be clarified by reference to actual programs used to direct the motions of robots.

## Industrial Robotic Products

In this section, a few examples of industrial robotic mechanisms will be given. Though it may be somewhat unfair to the many industrial robotic manufacturers, we will consider three main examples here.

There are more Unimation robots in use throughout the world (6,000 or more) than there are robots in use in this country. Joseph Engelberger is one of the industry's leaders and a most articulate spokesman on the future of robots. He is also president of Unimation. One of the robots he produces is the PUMA robot. This robot is driven by electrical motors which are necessary for precision work. GM was one of the first to use this robot on its shop floors. Each of its joints is controlled by a microprocessor (microcomputer chip).

Another robot used on the shop floor in assembly work is IBM's Rs-1 that can be controlled through the IBM PC through the use of the AML programming and robot control language.

Another model is the Intelledex, Model 605, that is not simply a mechanical arm like the above two. It can repeat to within .001 of an inch motions that it has been taught. This level of precision (and greater) is required for electronic assembly work. Using an integrated vision system, it can find its way to within .002 of an inch; it is programmable in BASIC, and it has the ability to set itself up on the factory floor. It can handle a variety of different weights and tasks.

The majority of industrial robots in use today are single-function, however. They are bolted to the floor, and have as a fundamental structure, the form shown in Figure 7:

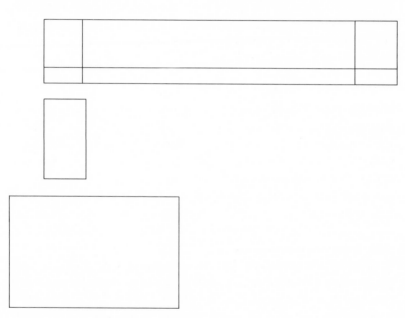

Figure 7.

## Personal Robots

Just in the last two or three years several manufacturers have produced kits or fully assembled personal robots that are general-purpose, maneuverable, programmable, quite capa-

91

ble given the early state of the technology, and have applications in a wide variety of places.[32] Available personal robots can be obtained for under $1,000, and those with greater capabilities may have prices ranging up to $6,000. Though the software for these mostly experimental robots is somewhat limited (see next chapter), the applications for them in the home, at school, and at work or play are really quite numerous, if someone has the patience and knowledge. At this stage of the game, personal robots are where personal computers were six or seven years ago—they are available, but they lack a pool of software, and all the bugs have not yet been worked out of them.

## Components Of Personal Robots

Most personal robots are built with sensory capabilities so that they may respond to the external environment. Using sonar transducers, photodiodes, tactile and obstruction sensors, light sensors, motion sensors, and sound sensors, personal robots are able (when the proper software is used) to map out "spaces" (for instance, a room) and to "learn" their way around in those spaces. Not only are instructions needed so that robots can do these things, they also need sensors of various kinds to be able to do them effectively. By being able to detect light and dark, the presence and absence of sound, the presence or absence of obstructions, a personal robot can perform any number of useful tasks.

By detecting light and darkness, it can be programmed (instructed) to do various and sundry things such as work during the day and "sleep" at night, turn the lights in a house

---

[32]Numerous articles are in existence with more always being written on personal robots. For example, see Daniel J. Ruby, "Computerized Personal Robots," *Popular Science*, May, 1983, p. 98, for a description of personal robots and their manufacturers. Manufacturers described include RB Robot Corporation's RB5X, Heath Company's HERO-1, Androbot Incorporated's B.O.B. and TOPO, and Robotics International's Genus.

on and off depending on the conditions, wake up household occupants who continue to sleep on through the daylight hours, and so on.

By being able to detect the presence and absence of sound, a robot can then be actuated by a human voice; it may respond in speech to voices that it hears; it can answer the door and welcome guests the way Albertus Magnus' robot was supposed to have done, before a fearful priest allegedly destroyed it; it can serve as a watchdog, sentry, guard; it can entertain by singing, and it can be useful as in helping children with spelling lessons, and so on.

As was true in the case of industrial robots and their sensors, the sensors used on personal robots—while quite remarkable—are really low-level sensors, as sensor technology has not yet developed very far.

Also still in its early stages of development is pattern recognition. Using available sensors, personal robots can detect still and moving objects (this is a form of perception, a form of differentiating the mass of sensations), and can determine the range of objects within its sensory environment.

In the space and military arenas, sensors of great power and versatility are available, though their cost is enormous. One of the military's radar networks, for example, costs hundreds of millions of dollars. It is so constructed that an object the size of a grapefruit can be "seen" and tracked from a distance of over 30,000 miles. Once mass uses are found for these sensor devices and have been declassified by the military, their prices should go down dramatically.

However, as already pointed out, sensors alone won't do the trick. The need is for sensors of the right type to be connected to software or interpretation hardware, and that the interface is at the right point. Human perceptual capabilities allow complex orientation and positioning in space, and the performance of many different sorts of task. We are able to orient ourselves, find our ways about, describe the environment, solve complex problems, dream, create, be lazy, identify objects and people, and detect and make judgements not just about still versus moving, but moving rates, speed, direction, and so on. Finding ways to duplicate these functions, not to mention all the higher functions involved with intelligence, is a major area of AI research.

In many ways, the problem lies in the technology. While it is true that the invention of chips made it possible to compress thousands of transistors onto a surface a fraction of the size of a transistor, that is still *not small enough*. The neurons of the human brain are considerably smaller and more versatile than the circuits of the chips that we have so far. It may be that once chips have been collapsed and placed into atoms themselves, and once organic chips have been developed, then we will be able to cram enough computing power into a reasonable amount of space and thereby somewhat duplicate the power of the neurons in the human brain. In short, we could most likely build a robot today that could sense and perceive with the power of a human brain, but that robot would have to be huge, the size of a football field, and thus not much help. Once we have developed atomic and organic chips it may be possible to duplicate the neurons in the human brain in some manageable amount of space, say the size of a human head. Until then, ingenious ways will have to be found to either duplicate or isomorph the sensory and perceptual capabilities.

Personal robots that can speak are common. Manufacturers have included phoneme speech synthesizers with several models that allow robots to "speak". This area of the technology is quite advanced. A robot could without difficulty be given the power to speak in any of the major languages, though "recognizing" and understanding such speech, i.e., when spoken to, is still a problem.

The power/drive system used in personal computers is mostly the electric motors and omnidirectional wheel-triads system. By using batteries, robots can work without direct and constant connection to an external power source, and by using rechargeable batteries they can detect when their batteries are running low, and in some cases can both tell a local human of their plight or can actually find an electrical outlet and can plug themselves in for recharging. Once solar power is added to the robot's power capability, then even this external dependency might be eliminated. In addition, once a suitable mechanism is found for locomotion, robots may be able to move around not just on level surfaces, but could be made to hop over obstructions, go down ropes, climb up ladders, and so on.

An arm-wrist tool is sometimes an option, sometimes a standard feature of personal robots. Through its use, a robot is able to pick up objects of a size and weight that depends on the strength of the gripper mechanism (roughly one or two pounds as of this writing, though much more powerful grippers are in existence). In one model, the arm-wrist tool has a capability of rotating up to 250 degrees and to rotate the wrist by 180 degrees.

Of course, it is far easier to do many tasks with two hands than with one. This step is one that I would like to see taken, even up to and including many arm-wrist tools of various kinds on one machine. There is no good reason why a robot should be limited to two arms like human beings. At Stanford, research has been underway for about a decade on two-armed robots. Though such robots are not currently developed by the R.B. Robot Corporation—the company with the most advanced "practical" personal robot—there are several manufacturers who have attached two arms to their personal robots.

As far as creativity in personal robots is concerned, that day is way ahead of us. As already noted, I think that day will await atomic and organic "chips" or electronic circuits that can mimic or that are isomorphic to human neurons. Create an organic or atomic neuron, and you can probably create a creative robot. We could probably create one now, but it would be too big.

## Examples of Personal Robotic Hardware

The earliest example of a personal robot took the form of a turtle or mouse that could be used experimentally. One example is the Terrapin Turtle that has been around for several years ever since it was produced by a student at the MIT AI laboratories. This turtle and related mechanisms like the mouse have been around since the early 1980s.

When we speak of mass-produced, generally available personal robots, however, we have to look to the three personal ("home") robots introduced to the market in 1983 and 1984.

HERO-1 The first commercial home robot was introduced in January, 1983, by the Heath Company. Over three feet high and weighing twice as much as the RB5X, the HERO-1 has similar capabilities. It can move and find its way around, and has an arm-gripper mechanism that can be used to fetch and pick up small objects. With its microphone and ultrasonic sensors, it can "hear" sounds and responds haltingly to these sounds. It is also able to talk and sing. In kit form, the HERO-1 costs around $1,500; assembled, it costs $2,500. Though it comes with an on-board computer that can be programmed in HEX (machine-level code), a number of peripherals manufacturers have developed interfaces so that it can be programmed in BASIC using any of the popular personal computers, like the Apple.

As of 1984, Heath had sold more than 10,000 of its robots, making it the leader in the field. There are now more Heath robots in the world than any other personal robot. Only the future will tell whether Heath will become the IBM of personal robots.

In 1984, Heath introduced the HERO Jr. which is not programmable, though it does have selectable "personality profiles", i.e., selectable preprogrammed functions. It costs $1,000.

RB5X. Two weeks after the introduction of the HERO-

1, RB Robot Corporation introduced the RB5X. Less than two feet tall and weighing 26 pounds, the RB5X can be equipped with an arm unit that can lift small objects up to one pound, sonar and bumper switches to enable it to find its way around a room, voice recognition, a vacuum cleaner attachment and a compass. In addition to these capabilities, the RB5X can "sing" and can play games. So far, around 2,000 units have been sold. The basic unit costs around $1,700. Peripherals like an arm, voice recognition, and vacuum cleaner attachments will double the price.

TOPO. Also introduced in 1983 was Androbot's TOPO prototype robot. Unlike the RB5X and HERO-1, the TOPO robot could only be programmed using an external computer. It was, in effect, a teleoperated robot. A new and improved version became available in 1984. The price for this unit is around $1,500.

B.O.B./xA. The B.O.B. robot, also from the Androbot Corporation, is the most advanced home robot now available. Introduced in 1984, its base price is $2,500 which is about what an RB5X or HERO-1 costs.

Unlike the RB5X and HERO-1 which come with limited on-board memory (16K in the case of HERO-1), the B.O.B./ xA comes with three megabytes of on-board memory and three on-board microprocessors (microcomputer central processing unit chips). Equipped with an arm-gripper mechanism, this robot can go to a specially developed robot-fridge, open the door, take out a beer and bring it to its master. It can also be used to wake someone, and its infrared sensors can be used to follow any selected object around, as, for example, in babysitting. It can also speak, answer the telephone, provide security alarms and so on.

Genus. This entry from Robotics International is the most expensive of the lot at $5,000. Its capabilities include the abilities to walk, talk, play games, sing, read and vacuum the floor.

Marvin Mark I. The Marvin Mark I is a product of Iowa

A Layman's Introduction to Robotics

Precision Robotics, Inc. It appeared on the market in 1984. For $6,000 one gets a personal/educational robot that is versatile and useful. The Marvin Mark I has its own on-board computer that is as powerful as many personal computers. This computer uses the popular CP/M operating system and is programmable in FORTH, which is one of the better languages for robotic programming. Like LISP, the preferred computer language of the AI community, FORTH is suitable not only for programming manipulatory activities of robots, but is very suitable for developing "expert systems," that is, software which mimics high-level intelligence in humans.

The Marvin Mark I has other attractive features. It is, for one thing, two-armed. Each of its six-axis arms can grip objects up to five pounds in weight, and both arms can be coordinated for the performance of complex two-armed activities.

HUBOT. Hubotics, Inc. introduced "HUBOT", the first home robot that is also a personal companion, educator, entertainer and sentry in 1984. Hubot can talk using a 1200 word vocabulary, can teach math and spelling, and has a built-in computer with 128K memory, display, keyboard, printer and programs. It is regarded as the "intelligent appliance of the 80's".

Others. Some home robots are available as kits to be put together by those who are interested in learning about robotic mechanisms directly (the HERO-1 is also available unassembled). In addition, there are about 100 companies that produce components for constructing robots for home or work. Depending upon the application, there are motors of different powers available for the drive mechanisms, solenoids and other switches and sensors available as sensing devices, various arm-gripper mechanisms, and, of course, a whole array of computer equipment for developing "intelligent" and programmable robots. Since the market for robot hobbyists is rather limited, we can expect that most robots used in the home will come in preassembled form.

# Robotic Mechanisms of the Future

Since we are in the very early stages of robotic technology, it is too early to tell what the machines in two or three years will be like. A lot depends on the entrepreneurial instincts of robotic engineers and pioneers, and a lot depends on how people generally respond to the availability of personal and industrial robots.

It is clear that now that American industry has been stung by Japanese productivity gains in the use of robots, that it will begin to apply advanced technology. The market for industrial robots will probably increase dramatically over the next few years as American industry moves forward. It is estimated that the tools in American manufacturing establishments are over 20 years old on the average, and seriously out of date. Retooling America using robotic tools is now a major drive, though the capital requirements of retooling are extremely costly.

With a small and selective market for industrial robots, manufacturers of these robots can afford to develop special-purpose, limited-task robots. But when the entire economy starts getting into robots, there will be a great need for general-purpose industrial robots that can adapt to a multitude of different environments and situations. I foresee the development in the future of versatile industrial robots that can be used in many different industries.

For personal robots, I see a similarly dramatic growth, delayed by a couple of years after the explosion of industrial robots. The company that can develop a multipurpose, versatile, adaptable and economic robot will revolutionize personal and social life in unimagineable ways. I could use a half-dozen good robots in my work, as housekeeper, personal secretary, general laborer, researcher, companion for children and animals when I'm away or busy, and general overseer and scheduler.

99

# A Layman's Introduction to Robotics

Seeing that the tools that humans use to take care of daily chores may not be suitable for robots to use, I foresee a robotic side-industry that will be engaged in the development of attachments and appliances that robots will be able to use—vacuum cleaners, ovens, washers and driers, TV sets, etc. Perhaps even transportation will be affected.

Other industries will arise to develop robots which can help people do a better job or to replace people in dangerous or hostile environments. We know that a good deal of work has gone into the problem of developing robots suitable for work in the hostile environment of outer space, but work is also under way to develop robots to do work in the hostile environments of earth.

Development of agricultural robots is already far advanced, for example. Robots are being developed which can distinguish between weeds and edible plants and thus can serve in checking weeds; others are being developed which can take soil samples, pick oranges, and grade apples. If Midwestern—agricultural—colleges and universities in Indiana (e.g., Purdue) or Missouri, Iowa and others have their way, robots for agricultural work will be prime areas of development in the years to come.

Other areas of human work are also being researched for replacement by robots. Areas where it is dangerous for people to work, for example as traffic conductors on crowded streets, can be taken over by robots. A Massachusetts firm is developing prison guard robots, while Australian robotics is focussing on sheep-shearing robots. Recently, the New York City Police Department awarded its "cop of the month" citation to a robot named RMI-3. RMI-3 helped end a 20-hour confrontation between police and gunmen in Elmira, N.Y. by using its arm to open a door, entering the apartment where the gunmen were holed up, and using its camera "eyes" to alert the police when it was safe to enter the apartment.

100

# Illustrations

Two examples of single-purpose 'promotional' robots. These are not true robots. The robot on the right can be remotely controlled to move objects placed in its tray (e.g., drinks) from one point to another in a room. These robots can be used for elementary education and instruction in robotics, and for low-level simple function tasks. (*Picture courtesy of 21st Century Robotics, Norcross, Georgia*).

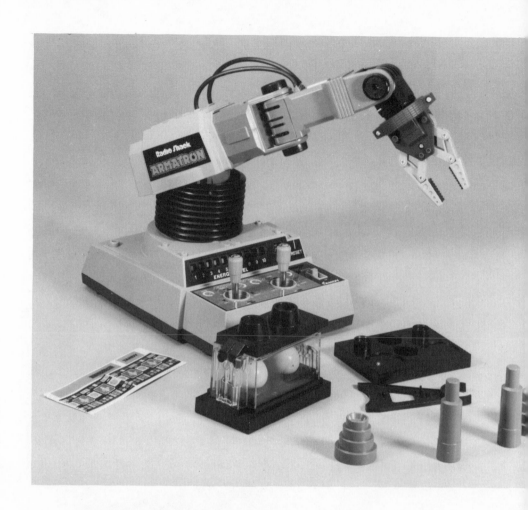

Another example of a low-level, single-purpose, robot is this ARMATRON (tm) from Radio Shack. This very inexpensive robot can be used by children in learning 'pick and place' tasks using joysticks for 'program' control. (*Photo courtesy of Radio Shack, A division of Tandy Corporation*).

Pictured below is the HEATH Company's HERO-1, the first program-mable personal/home robot on the market. Considered an educational robot, HERO-1 is shown here with its optional remote radio controlled keyboard which allows complete operation from a distance (held by the boy). Either the remote control or the robot itself can be programmed from a large computer (shown on the desk). (*Photo courtesy of Heath Company, Benton Harbor, Michigan*).

Looking very much like R2D2 of Star Wars, RB Robot's RB5X Robot entered the personal computer marketplace two weeks after the HERO-1. Here it is shown holding several of its electronic options in its grip. (*Photo courtesy of RB Robot Corporation, Golden, Colorado*).

Billed as a functional, home robot, the HERO Jr. was introduced early in 1984. Though non-programmable, the HERO Jr. has several built-in 'personality' modules which can be selected by the user. These pre-programmed 'modes of behavior' include the explorer mode, the make-the-guests-feel-at-home party mode, and the home security mode. (*Photo courtesy of Heath Company, Benton Harbor, Michigan*).

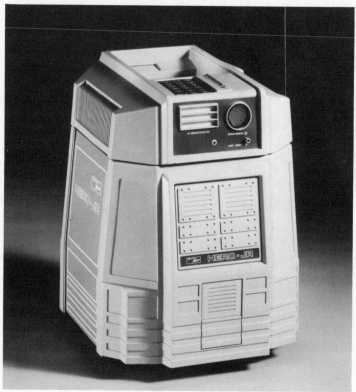

106

107

Pictured opposite is the ZYLATRON robot developed by Otis Observatory. This illustration serves to strip away the 'outside' and show the inside of robots. As seen here, a robot is a combination of standard mechanical components (wheels, drive belt) and electronic components (wires, batteries, etc.). This model can be programmed in BASIC, has built-in artificial intelligence, can speak in any language, has sound and light-detecting sensors, and can serve as a somewhat bulky household companion. (*Photo courtesy of Otis Observatory, Aberdeen, SD*).

Harvard Associates' Turtle Tot is a programmable robot which moves, uses touch sensors to feel its environment, draws with a pen, blinks its eyes, and can even talk with the speech option. Can be programmed from most of the better selling micros, e.g., APPLE and IBM. Originally developed in the AI program at MIT, the Turtle Tot can serve as an affordable 'pet' for those people interested in moving from personal computing into robotics. (*Photo courtesy of Harvard Associates, Somerville, Massachusetts*).

108

109

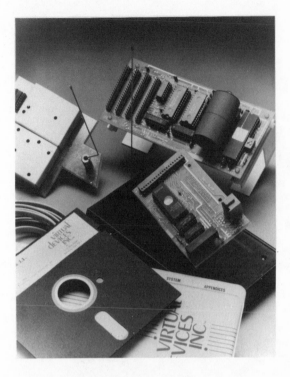

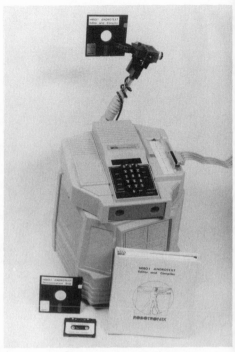

As in the case of personal computers, one of the fastest growing industries is the robot peripheral support industry which produces the add-on optional peripherals which help to make robots more useful. Picture on top-left shows the electronic boards which make up the MENOS robot control language and programming development system for the HERO-1. At the top-right is the HERO-1 gripping a floppy disk of the Robotronix ANDRO-TEXT editor and compiler, a BASIC-like language for personal computer programming. The two bottom photos show the voice command system and the APPLE-HERO communicator from Arctec Systems. (*Photos courtesy of Virtual Devices, Inc., Bethesda, Maryland; Robotronix, Inc., Los Alamos, New Mexico; and Arctec Systems, Columbia, MD*).

The top photo shows a PRAB G-26 model industrial robot performing a standard 'pick and place' operation on a tire rim. The bottom photo shows another PRAB robot working on the assembly line. (*Photos courtesy of Prab Robots, Inc., Kalamazoo, Michigan*).

Industrial robots have been found to be more efficient in handling 'dangerous' materials like glass. Here we see a Unimate 2000 from Unimation, Inc., the maker of the greater number of industrial robots in use throughout the world, performing a glasshandling chore. (*Photo courtesy of Unimation, Inc., Danbury, Connecticut*).

112

These two pictures show the Unimate 2000 performing die-casting chores. (*Photos courtesy of Unimation, Inc., Danbury, Connecticut*).

113

Here we see the Unimate 4000 welding at one of Chrysler's plants. Studies have shown that the welding is more accurate when done by robots, with less waste, and above all less hazardous duty for human beings who can fall victim to serious industrial accidents when welding on the assembly line. (*Photo courtesy of Unimation, Inc., Danbury, Connecticut*).

114

## Photo Credits

Once again, we would like to express our thanks to the following firms for their permission to reproduce the photographs in this section.

| | |
|---|---|
| Illustration 1 | **21st Century Robotics** |
| Illustration 2 | **Radio Shack** |
| Illustrations 3 & 4 | **Heath Company** |
| Illustration 5 | **RB Robot Corporation** |
| Illustration 6 | **Harvard Associates** |
| Illustration 7 | **Otis Observatory** |
| Illustration 8a | **Virtual Devices, Inc.** |
| Illustration 8b | **Robotronix, Inc.** |
| Illustration 8c & d | **Arctec Systems** |
| Illustration 9 a&b | **Prab Robots, Inc.** |
| Illustrations 10, 11, & 12 | **Unimation, Inc.** |

# Software
## of
# Robotics   5

---

## Introduction

By software of robotics, several different things are meant. Software is, first of all, soft in the sense of being malleable and changeable, e.g., reprogrammable. It is software also in the sense of being nonphysical, of being a mental construct on a computer created by a programming "mind". It is part of the "wares" or the commodities of the information processing part of robots. Malleable and reprogrammable wares of robotics are programs, similar to programs that run on computers. These programs all serve one principal purpose— to specify the set of procedures or algorithms involved in *doing* anything at all. In order to do, robots need software.

Programs are coded into a machine through the use of a programming language. Some of the programming languages used in robotics are simply elaborations on already existing languages for programming computers, while other languages are new developments. For example, TINY BASIC, a subset of BASIC, is used by RB5X. LISP, which is widely used in AI work, is also used as a robot operating language. On the other hand, IBM has a robot programming language that it calls AML (A Manufacturing Language). This language is considered by IBM to be the most advanced robot programming and computer programming language around.

There are at least three levels of software (see Figure 8). There is software that is designed to *do* something, for example software directing a robot in the performance of daily chores, directing it how to cook a duck, telling it how to open a door. This is what we can call application software. It is software that directs a robot to do something we want done: play with the children, empty the garbage, etc.

On a higher level is "control" software, that is, software directing the robot on how it is to interrelate all of its different parts, how it is to interrelate its voice with its action, movements, hearing, vision, and so on. While application software

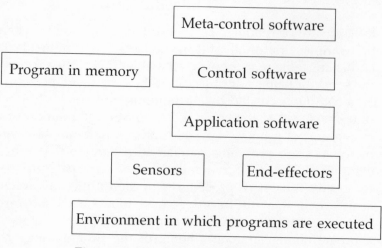

Figure 8. Types of robotic software.

118

is analogous to a human moving his hands and legs, control software is analogous to the brain where control is exerted over these movements.

On a still higher level of generality are various meta-control rules that have been proposed for robotics, among the most famous being Asimov's three laws of robotics. These laws supposedly govern the relationship between the control software, the application software, the hardware that effected the required actions, and the wider environment of human society. This level of software is analogous to the ideas that issue forth from the cerebral cortex, the front of the brain, the center of thought where all the "higher" functions like language, thought, imagination and so on take place.

We have already considered in Chapters 3 and 4 the environment, sensors, and end-effectors, so here we will consider mainly the software types. The five topics to be considered in this chapter are:

the nature of software algorithms

programming languages

application software

control software

meta-control software.

## Software Algorithms

An item of software, one could say, is the statement (in some language understandable to a computer) of a procedure to *do* something. It is the statement of the steps that are to be followed to accomplish some task, whether mental (such as adding two numbers) or physical (such as moving between two points).

Generally a procedure is a *way* of doing something, a

119

means, a mode, a method, a style, system, technique, process, course, line of action, mode of operation, routine. It is in short, a statement of *how* something is done or is to be done.

There are two sorts of algorithms, two sorts of ways, that can be followed. Only one of those two ways is appropriate for robots. One way has the following characteristics: it is entirely infinite, indefinite, and unrepeatable. This is the method of indefinite infinity. To accomplish any goal, to do anything, one is urged to accept the possibility of taking an infinite amount of time and an infinite number of steps to do something; on the other hand, no one thing, or no definite set of steps, can guarantee any result. This method cannot work with computers or robots or any other being with a task to perform.

The other sort of algorithm has a contrary set of characteristics: it is finite, definite, repeatable, and public. It may be possible for a bunch of monkeys, given an infinite amount of time, to write the works of Shakespeare,[33] but if you want to be sure that the works of Shakespeare get written, then make sure that there is a finite person (like Shakespeare), with a finite amount of time to do his work, with a definite set of ideas about which he wishes to write, and that the works can be repeated in public.

This second algorithmic sense is the sense used here. Software items describe (in a programming language) the definite, finite, and repeatable set of steps a robot has to go through to accomplish any task. For example, a common and simple task to assign a personal robot is to map out a room. Giving the robot, first of all, a grid to use, we instruct it to move forward (or backward, or sideways) one step. If there is an obstruction, step back, and try a different direction. If there is no obstruction, then continue going forward, or turn in some other direction, and so on. Using this simple method,

[33]This has been challenged as an impossibility.

120

a robot can map out on a grid the location of obstructions and free paths.

## Programming Methods and Languages

To state an algorithm, it is necessary, of course, to be able to write it down and express it logically, but it is also necessary, if it is to be executed by a machine, to state the algorithm in terms of some programming language. Since a program is a statement of an algorithm—that is, of how something is to be done—the programming language used to instruct a robot must have commands that are related to movement within the confines of "real" space, not the space or the illusion of space created on CRT screens. The programming language used must be one that can instruct a robot to move in certain directions and to position itself at various points, and it must also be able to direct the movement of the end-effector, the arm-wrist tool that does the work.

There are three methods used in the development of software for industrial or personal robots:

One method or language is really prelinguistic in that it relies mainly on "showing" a robot what to do.

Another method is really a futuristic wish list.

A third method is the method used in the computing industry: programming robots using an explicit programming language.

We will deal with each of these in turn.

### Prelinguistic Programming

The method of programming a robot that is most common in industrial robotics is a method that is prelinguistic in the sense that it relies less on explicit statements to "tell" a robot

121

what to do, as it does on showing a robot what to do. Using an accessory called a "teaching pendant" a robot is shown what it is to do. Thus, this method could be called the "show me" method of programming.

This method consists in showing a robot what to do. A robot that uses this method of programming need not have a sophisticated computer on board. Only a few switches and buttons need to be built into the robot. The problem with this method is that while it is the easiest way to "program" a robot, it is nonetheless limited to simple actions (movements) and simple programs.

This method works in three principal steps:

First, a button is pressed indicating that the robot is in *teach* mode.

Second, the robot is shown what it is to do using the teaching pendant. The memory of the robot is told to store the action it is shown.

Third, another button is pressed to indicate that the robot is to go into "action" mode and is to execute the selected action.

A "program" in this mode of programming would look like this:

1. First, one would indicate via a push button that the goal or point of the program is such and such. One would, for example, press the GOAL button, then a letter of the alphabet indicating the "name" for the goal.

2. One would then press another button indicating TEACH mode. This would get the robot to act in a particular way, i.e., to be receptive to being taught a procedure.

3. One would then use the teaching pendant to guide the robot manipulator or end-effector through the paces required to achieve goal A.

4. Once the procedure had been indicated, one would press another button to instruct the robot to store the steps shown in its memory system on disk or tape or an external storage device.

5. Once one is ready to have the robot do action A, one would press another button indicating that the robot is to execute action A or action B or whatever other actions may have been stored in memory.

This method of programming is used on the Unimation Unimate robot using the VAL programming language. Another language, RAIL, is used for the same purpose on the Automatrix AID 800 manipulator.

This method of programming is popular in spray painting applications, for example. A skilled painter would use the pendant to show the robot what to do, then when instructed it would duplicate the actions indicated. Most industrial robots use this method, as do the personal robots like the RB5X and the HERO-1.


## Comprehensive Programming

A second method of programming a robot is the comprehensive method. This method is known in the industry under various names, such as the "world modelling" method. Using a variety of CAD/CAM methods, the steps that a robot is to go through to accomplish a task are developed using computer-assisted design methods. Instead of "showing" a robot what it is to do, this method simulates a robot procedure using three-dimensional geometrical models. By simulating robot actions on a screen using a geometry based on

# A Layman's Introduction to Robotics

Cartesian coordinates, each step can be indicated by using the model.

Though complex assembly processes can be developed using this method, one of its problems is that it assumes that the way the "real" world operates is the way the "model" operates. As such, then, there are no sensors to indicate that there is something amiss between model and reality, so that corrections cannot be made. This is a problem found on IBM's Autopass language.

One of the chief features of this method is that assuming that the model matches the real world, then complex tasks can be programmed and accomplished. Built into the robot are complex utility programs that can execute complex tasks with a small amount of explicit programming statements.

For example, we could state a program for a robot using this method in a simple program:

```
10 PUT BOLT IN NUT SO THAT NUT AND BOLT LINE UP
20 CHECK SENSOR A
30 IF A=1 THEN SWIVEL GRIP 360 DEGREES: GOTO 20
40 STOP
```

This simple program accomplishes the goal of getting a robot to screw a bolt into a nut. The first line of the program is in "English"; there are complex routines to interpret the English correctly.

The next step is to check a sensor, for example, a sensor to check when the bolt cannot be tightened further. If sensor A indicates that the bolt can be screwed further (A = 1), then the robot would swivel its gripper by 360 degrees, i.e., it would tighten the bolt another turn. Each time it performs a turn, it checks the sensor.

Once the sensor indicates that A = 0, the robot would know that the screw has been tightened. The next step in the program indicates that it is then to STOP doing what it was doing and await further instructions from the operator.

124

**Explicit Programming**

This method of programming a robot is similar to the method most commonly used in the computer industry. It consists of a robot and computer programming language. In most cases, a high-level programming language is used along with a suitable subset of motion and manipulation commands.

The focus of this method of programming is on the end-effector or manipulator, and on the instructions of what the end-effector is to do each step of the way. The robot is not shown what to do, nor are complex geometrical models developed. Instead, the robot is instructed each step of the way on what it is to do in any application.

There are a few explicit programming languages used in robotics:

There is SIGLA developed by Olivetti of Italy for its Sigma robots. This language developed out of two languages developed by Stanford University, one called WAVE (developed between 1970 and 1974) and the other Al (developed in 1974). Both are based on the ALGOL computer programming language.

There is HELP which is used on GE's Allegro (tm) robot.

There is TINY BASIC used on the RB5X personal robot.

There is IBM's AML language which was recently adapted for the IBM PC.

There are other high-level languages like Pascal, C, and Forth that are used by various hobby-oriented institutions.

There is ANDROTEXT of Robotronix, Inc. which is a high-level language for personal robots.

## Application Software

Application software is software developed so robots can *do* things within the range of their end-effectors, sensors and power/drive mechanisms. Since the applications are so different in the two cases, we'll consider personal and industrial application software separately.

### Personal Robots

There are three or four ways in which a personal robot can be programmed. One way is through an external program. Such a program could be one that could not "fit" on a robot or in its memory, so in effect the robot takes on the form of a "dumb terminal". Androbot's TOPO, introduced in 1983 in prototype, is an example of a nonautonomous, externally programmed robot.

A second way of programming a robot is to have on-board programming tools (a built-in computer and terminal, in other words). Most of the available personal (or home) robots use this second method, such as RB5X, B.O.B. and Genus.

Because personal robots are in such an early stage of development, there is very little demand for applications software. The pioneers and experimentalists are busy thinking up and developing software for their own use. As of the end of 1984, at the most 10,000 personal robots are expected to be in use. Once around 100,000 personal robots are in existence, however, we can expect to have a great deal of software.

As of now, as we saw in the previous chapter, personal robots of various sorts can be programmed to sing, talk (up to 150 words), play games, retrieve objects on command, watch for burglars, wake people up, answer the telephone, vacuum the floors, and so on. But can they do windows?

This may not be so far off. Right now, robots are being

used in educational software programs (e.g., RB Robot Corporation), and programs to do useful things are being transferred from fixed and unimodal robots into multifunctional robots. In fact, Androbot's B.O.B./xA even does windows.

## Industrial Application Software

The application software—or software-hardware combinations—available for industrial robots is quite extensive even considering the somewhat limited capabilities of industrial robots.

There are robot systems (hardware and software combinations) at work throughout the world, but mainly in Japan (where 70 percent of all industrial robots presently exist) and other industrial nations. These robots are performing tasks like:

forge work

loading and unloading materials

tending machines

pick and place

linear welding

thermovision scanning

electronic assembly (e.g., assembly of printed boards)

arc welding

quality inspections.

As capabilities in programming languages, pattern recognition, sensors and motor-drive systems improve, the range of applications should increase dramatically. The manufacturing industries, presently reeling from Japanese productivity gains using robots, can be expected to be one of the

A Layman's Introduction to Robotics

most automated industries by the end of the century. Already at such plants as GM, Chrysler, and among the Japanese automakers, robots are in wide use. In some cases, as high as 98 percent, robots have taken over all welding and painting chores on the factory floor.

**Example Of A Programmed Application**

So as to give the reader a sense for an explicit program for robot execution, we will use a paper model of a robot programming language. Instead of using one of the existing languages belonging to a specific vendor, the BASIC programming language so popular in use on personal computers can be adopted for illustration purposes here.

The problem to be demonstrated is one that can be solved by either industrial or personal robots. The objective is to move object A from point 1, say a table or feeder, to point B, say a pallet or a box. In industry, this would be a palletizing operation; at home, it might be an operation to pick up crayons from the floor.

For purposes of the illustration, assume that there are ten objects to be moved.

```
5  REM PICK UP PROGRAM
10 REM MAIN PROGRAM
20 FOR I = 1 TO 10
30 GOSUB 100
40 NEXT I
50 STOP
```

This is an example of a main program programmed in BASIC. First, we use a remark (REM) line to note the name of the program. Then we give the instruction that the following procedure is to be done ten times; once it has been done ten times, then the program should terminate. The action that is to be done ten times is in the subprogram (or subroutine) that is listed next.

128

```
100  REM MAIN SUBROUTINE
110  REM FIRST MOVE GRIP TO POINT A
120  MOVE GRIP UP TO 10 (Z, 10)
130  MOVE GRIP TO COORDINATES X(3.5), Y (2.07)
140  MOVE GRIP DOWN 1 (Z, 1)
150  REM NOW SEE IF ANYTHING'S THERE
160  CHECK SENSOR INFRAREDBEAM B
170  IF B = 0 THEN GOTO 240
180  REM OTHERWISE DO THE FOLLOWING
190  CLOSE GRIP
200  MOVE GRIP UP 1 (Z, 1)
210  MOVE GRIP TO COORDINATES X(13.3), Y (2.1)
220  MOVE GRIP DOWN 1 (Z, 1)
230  RETURN: REM TO NEXT I
240  PRINT "THERE'S NOTHING THERE. "
250  STOP
```

This program gives the reader a sense for the principal elements of a robot application program. Some additional commands that may be found in a particular robot programming language include:

Robot Movement control commands like CLOSE or OPEN GRIP, MOVE TO, MOVE STRAIGHT, MOVE IN AN ARC, DEPART, APPROACH, ACCELERATE, DELAY, WAIT, etc.

Typical structures such as BEGIN . . . END, WHILE . . . DO, LOOP, etc.

The usual variety of arithmetic and logical operators like GT, GE, AND, OR, etc.

## Control Software

The main function of control software is to control, monitor and direct all the functions that a robot performs. Control software operates usually in two modes (as do thermostats)— automatic control and human override.

129

# A Layman's Introduction to Robotics

A robot control program is similar to control programs (e.g., monitors or CP/M) on computers. A control program monitors and governs the occurrence of events in proper sequence and timing. Control programs govern and control accessibility to external databases, monitor all sensors periodically, and respond to the environment by interpreting the situation and by selecting the appropriate application software (stored in memory) for use at any one time.

Controllers can, of course, be many things and have many levels of complexity. The more complex a control program, the more capable the robot. A control program for robots programmable in the prelinguistic, "show me" style of programming would coordinate the few pushbuttons and toggle switches (TEACH MODE, DO MODE, STORE, etc.) with each other. In this programming mode, a control program would (1) store data as instructed, (2) initiate and terminate movements as instructed, and (3) interface to the user and other external devices.

A robot control program on the level of an explicitly programmed robot would be a more sophisticated version of the previous type. It would coordinate storage, communication modes, devices and peripherals just as a computer control program does.

Control programs for "comprehensive" or "world model" programming would be even more sophisticated still. For one thing, instead of using *algebra* as the control characteristic, *geometry* would be used. That is, all digital computers use an architecture and programming languages and styles based on or modelled after algebraic systems, linear systems of sequential instructions. World model programs are, on the contrary, based on a geometrical style of programming where networks and simultaneous, rather than serial, instruction sets are used. Control programs on this level would not only coordinate external devices, but would have to coordinate and be able to call upon very complex programs that could

130

make calculations related to volume calculations, surface area calculations, etc. which are used in engineering disciplines.

## Meta-Control Software

While a control program per se controls and monitors the behavior of robot actions and application procedures, meta-control programs are those that monitor whole classes of actions, all controls, and all automatic and override operations with the aim of determining at each step whether the proposed actions are appropriate or not. Meta-control programs are what we call "moral" and "ethical" principles.

Moral principles do not tell us how to do things, what procedures to use, but rather what we should not do, plan to do, create a program to do and so forth. Similarly, meta-control programs—where they exist—tell a robot not how to carry out its tasks, but what in the final analysis it must or must not do under any circumstances.

In his 1950 book, *I, Robot*, Isaac Asimov proposed three meta-control rules that have been widely touted and repeated, and are regarded as almost sacrosanct.[34] These three laws of robotics were to be *immutably* wired into the positronic brains of Asimov's race of robots. These robots are known as droids, that is, as robots that are not evil and will not hurt human beings.

To paraphrase, these three laws are:

1.  No robot may injure a human being directly, nor may it allow a human being to be injured because of a failure to intervene directly.

---

[34]For example, Alvin Toffler supports the idea that these three laws must be programmed into robots, though he shows little if any awareness of what these laws really involve. Similarly, Joseph Engelberger believes that the three laws must be programmed into robots.

2. Robots must obey human orders except where it conflicts with the first rule.

3. A robot must preserve itself, except where such preservation would result in violations of laws 1 and 2.

These three laws will not stand up as adequate when subjected to analysis. I do not at present have any better alternatives, but I do know that these three will not work.

## Rule One—Do Not Injure A Human

No robot shall either injure a human being, or cause a human being to be injured through its neglect.

Let's first note that this "law" of robotics is equivalent to the usual laws in religious circles—such as the Golden Rule—to the effect that people should love one another. Whereas, however, the Golden Rule is not followed, and is not expected to be followed by human beings, the first law of robotics is supposed to be one that *cannot be violated in any way* by a robot. A robot is expected to follow the rule all of the time, and on all occasions, regularly and without any exceptions.

A very interesting inversion has occurred here. Whereas God created humans in His image, but on a lower level of imperfection, Asimov would have us create perfect beings that are *unlike humans* and like God in being regular and reliable. The first law of robotics, the first law governing a creation of the created human being, is not a law for a "third-level" being, but for the original creator itself, a first-level law.[35]

---

[35]To require of our creation fail-safe behavior, something of which none of us is capable, is to require that our creations should be "higher" in development than we are, an improvement on the normal behavior of the human species. This is a curious theological or cosmological perspective. It is even more curious as to how we are to implement this idea in a computer program. What we need to do, in effect, is to program the robot with a "calculus of morality," something that we have not been able to do for humans.

132

To require of a third-level creation behavior appropriate to a first-level being[36] is unrealistic on a practical level, and is a "domain error" or category mistake on another level of logic. The first law of robotics is logically and practically flawed. It applies to one domain what is really applicable to another domain. This is like complaining that babies are flawed because they can't speak the day they are born.

Since, as Huxley noted many years ago, it takes only one little fact (or counter-example) to destroy any beautiful theory, let us consider one such counter-example to the first law.

Suppose a robber is coming at me with a gun with the *potential* to inflict injury on me. By the second law, I command my robot to injure the approaching gunslinger before he injures me. My robot refuses on the grounds that it is not allowed to injure a human being. We need another law, to distinguish between good and evil, and to know what loyalty to an "owner" means. "My" robot should obey me, not another. My personal robot should act according to my instructions, not according to a law that it cannot apply and which, if tried, may result in a breakdown.

However, the second part of the first law states that the robot must not allow injury to come to another through its inaction. So, sensing that I would be injured, it felt a compulsion to prevent my being injured, yet it could not injure another to prevent my being injured. Locked in that mortal conflict over insufficient instructions, I go down at the hand of a gun, and my robot can still do nothing to the robber or his gun but must be himself uninjured and must obey the (now) murderer and robber.

Our conclusion is that Asimov's first rule will be unreliable precisely because of trying to make a robot follow a totally consistent rule.

---

[36]This does not entail that either of these beings exist, only that if they did, then such and such would follow . . .

## Rule Two—Obey Humans Except . . .

The second rule is that robots should do what humans tell them to do, except when what they tell robots to do violates the first rule. So, with a robber coming at me, I tell my robot to grab his legs or trip him. My robot springs to obey only to be tugged back by the first law that prevents him from injuring a human being (as may happen if he is tripped), but on the other hand my robot can't just stand there as he is commanded not to allow me or any other human—including the robber—to be injured. Lacking sufficient instructions, my trusty robot collapses in a sobbing heap of indecision.

A trusty "dumb" dog without sophisticated rules would be a better protector than my "ethical" robot. Thus, whereas the first law suggests that a robot would need a "self," the second law presupposes emotions and feelings like loyalty and ownership.

## Rule Three—Preserve Thyself Except . . .

I command my robot to throw himself in front of the gun to prevent the bullet from hitting me. "Are you kidding?" says the third law of self-preservation. "No, I'm not kidding," says the second law of obedience to human commands. "You've got to do something, but there's nothing you can do," says the first law of noninjury.

If seriously proposed as laws of human behavior, the three laws of robotics would soon spawn the collapse of coordinating functions. As laws for the behavior of any beings at all, the three laws of robotics are inadequate and unworkable, and should be scrapped.

# Implications of Robotics **6**

## Introduction

Robotic implications for the economy, society, politics, and life in general have received increasingly urgent attention in the media. The issues of the fate of the manufacturing industry and of the role of robotics in saving this industry from ruin are of great concern. These are problems concerning economics, rights, a new form of species, organic robots, the concept of personhood, and the idea of control over humans.

For example, there is the fear of the loss of one's job on the one hand, and on the other hand there is a welcome given to the new technology. As far back as we can trace in history, we find these two pervasive tendencies on the part of people: fear and loathing of new technologies, and openness to change and to the opportunities offered by new technologies. Given that new technologies have inevitably benefitted humanity as a whole, it would appear that the issue between those who fear and those who welcome technology is really a difference between egoism and the good

135

of the whole society. Put another way, those who fear new technologies are apparently concerned about their own selves while ignoring the benefits that accrue to the entire society as a result of the introduction of new technologies.

Another closely related issue is the legitimate concern of what is to be done for workers who are displaced by robots or by other new technologies. If new technology in fact displaces certain individuals and causes them harm and hardships, and if new technologies give life to the economy as a whole, then it would make sense to have the economy as a whole provide the means for retraining such individuals. In some American companies involved in a move into robotics, workers have been guaranteed that they will not be displaced, but will be retrained.

Another issue concerns the integration of workers into the decision-making aspects involved in integrating change and new technology into the workplace. People who are kept in the dark about their inevitable displacement by machines will inevitably develop resentments and suspicions about the new technology and will thus act as hindrances. On the other hand, workers allowed to participate and share in the decisions involved in introducing new technology may come to support rather than oppose the venture.

Another issue of concern is that involved in the psychological problems of machine and human interaction. Though there are psychological problems connected with any physical or mental technology, a technology like robotics that involves a fusion of mental and physical technologies and that involves the development of a complement to human powers is likely to cause even more problems.

In the area of personal robots, as distinct from industrial robots, there are various other problems that need to be considered in the society at large. One problem is related to the phenomenon of "computer widows" where computers have been brought into the home. In many cases, people get so involved in the new technology of computing that they become fanatical about it, so much so that they forget or ignore people as a form of contact. One might well wonder what will happen when people can relate to isomorphs of themselves, to machines with human-like qualities and abilities.

In addition, there is the whole issue raised originally by Capek's play, namely, what are humans to do once robots take over? Will robots turn us into their "slaves" in the sense that we will become so dependent on them that we will become incapable of fending for ourselves?

I share these concerns, and would like to consider some of them here.

What are the implications of robotics for the information society?

What impact may robots have on the manufacturing industries?

What are some of the social and economic impacts of robots?

What are the implications of robotics for the meaning of human life?

## Robotics in the Information Society

The processes involved in information are to gather it, enter it, store it, process it, display it, retrieve it, and so on.

At the present time, in most cases, people gather the data and people enter it; computers store, process, display, and retrieve that data. One of the uses to which computers and other automatic devices like robots will likely be put to use is in the development of ways to gather and enter data into computers.

Gathering data is an activity that has gotten human beings into both pleasant and unpleasant situations. Researchers,

137

spies, explorers, adventurers, pioneers, and sensors are all in the data gathering business. There are, however, many places human beings have never been, many regions are unexplored, many oceans have never been mapped or explored. Every so often, new and unknown creatures are discovered where they are not supposed to be. (Recently, scientists discovered a form of marine life that lives on deep ocean bottoms, and contrary to the then-known "laws" (regularities) of biology does not require sunlight to live, something every other being on earth is alleged to need.)[37]

Both on Earth, and far beyond it, there are places and events that need to be explored and that can be explored and researched using robots of various kinds. Robots could probe where it is unsafe for fragile humans. A robot could plunge into the sun, another could enter the holes in the ocean bottoms where the hot springs flow, others still could explore Loch Ness for Nessie, Lake Champlain for Champie, Washington State for the Sasquatch, and the regions around New Guinea for the legendary mermaid (known locally as the Ri, allegedly real animals.)[38]

John von Neumann's theory of self-reproducing automata is based on the idea that a real robot should be able to be given the instructions on how to manufacture the parts and assemble a robot like itself. This is a powerful concept. Von Neumann himself, and after him scientific speculators like Gerald O'Neill of Princeton, have imagined spectacular

[37]See the newsletter of the International Society for Cryptozoology, Vol. 1, No. 4, Winter, 1982, p. 8. These creatures do not depend on the sun for energy. This is supposed to be unique to the earth. The area is at 8,500 feet, 150 miles south of Baja California, at latitude 21 degrees north. Here, feeding off hot springs are enormous crabs, snails, worms, and other creatures new to marine sciences.

[38]International Society for Cryptozoology, P.O. Box 43070, Tucson, AZ 85733. O'Neill is the leader in this area of speculative research. Any of his books is worth reading, for he is at ease with the world and moves easily through it. He passes easily from "I had my doubts about horseback riding . . . " to speculations about robots being used throughout the universe.

uses for such robots. The common theme to these ideas is that one could send a robot of the self-reproducing variety to some distant outpost of the universe. There, one robot would proceed to duplicate itself a certain number of times in order to get "helpers" for its programmed tasks, which might include mining, or the creation of habitable structures, or something similar.

Though space is fascinating, there is still a great deal that we know nothing about on Earth, and under it. One or more robots could go on the track of animals that have not yet been discovered or classified. They could also be made to tunnel to the center of the Earth, perhaps beginning their digging at some deep California fault. If giant crabs, worms and other seagoing invertebrates and vertebrates have only recently been discovered to exist in what even at 8,500 feet is still open water, may there not be many other wonders of nature that could exist completely underground drawing *their* energy from the heat of the earth?

Robots could, of course, also be given the chore of data entry, the tedious business of entering items of data into computer data banks (a process prone to error when done by humans).

The information society is today dominated by data processing devices. The entrance of industrial and personal robots on the scene will involve certain changes of emphasis in these industries.

We can distinguish four principal emphases in the DP industry, four sorts of services and tools provided by them to the general economy:

1. Networks provided by the telecommunications side of the industry.

2. Databases storing different types of information and providing access thereto.

3. Programming languages and operating systems.

4. Application programs and software products.

In each of these areas, robots will introduce fundamental changes and expansions.


## Production and Robotics

Twenty-five percent of the labor force is involved in manufacturing. The overall profit level in this industry is 3.6 percent and is declining. Currently, an autoworker makes $24.00 per hour. The cost of robots, on the other hand, is $6.00 an hour for capital and maintenance costs. The principal economic drive for using robots is to reduce labor costs. GM alone plans to have 20,000 robots in use by 1990 and to spend $1 billion on this venture, while the entire economy can be expected to have anywhere from 150,000 to 500,000 robots in the US by the early 1990s or thereabouts.[39]

As we saw in an earlier chapter, industrial robots are quite versatile given the early stage of robotics, yet they are mainly lacking in external sensors that could be used to have them interact with their environment and respond to problems and events within it in intelligent ways.

But the really interesting implication of robotics is not so much over whether or when robots having such and such capabilities will appear on the scene, but rather *what* the effect will be of having robots doing a majority or even all of the production of goods for a society. One implication is that provided the goods are designed well and are put together using appropriate and long-lasting materials, then products are likely to be standardized and identical since robots are

[39]See no. 54, p. F-1, 1983 *Daily Labor Report*.

not likely to be given "creative" instructions and assembly procedures.[40]

Another implication of robots in charge of production is that the processes of making and assembling products will eventually come to be done mainly or entirely by robots. This leaves for human beings the two major tasks of control and policy-setting, and of marketing and selling the goods that are produced.

Now, if a general-purpose robot is one day able to reproduce itself, then it should also be able to construct just about any other product on earth. One could drive up into the Rocky Mountains and have one's robot eat up a few rocks and boulders on its way to producing various and sundry items that an individual might want to have. Instead of going to a store or putting in an order to a factory, one would simply command one's robot. But if the individual can have his robot produce whatever he needs, then of what use would be a marketplace or stores, or so many other production-related activities in the economy?

If general-purpose robots soon come into existence, then what happens to human work and human productivity generally? What do human beings do with themselves in such a situation? The last section of this chapter takes up this question.

## Robots and Robotics

The age of robotics is upon us and we're not prepared. As an example, YAMAZAKI of Japan has set up a factory using 65 computer-controlled machine tools, 34 robots, and 210

---

[40]This is false as computers add to rather than detract from individualization. One recent example where the computer has provided individualization that could not be achieved in earlier times is in the case of Coleco's rag doll where thousands were created in individualized patterns for people in the Christmas season of 1983. In fact, computerization gives the manufacturing industry the capability of customizing every one of its products.

# A Layman's Introduction to Robotics

## Table 13. Economics of robots.

| Method | Tools | Robots | People | Cost/hour |
|---|---|---|---|---|
| Nonautomated | $0 | $0 | 2500 | $16/hour |
| Automated | $65,000 | $35,000 | 210 | $4/hour |

people all linked together through corporate headquarters situated elsewhere.[41] The use of the robots and other automated tools required only 210 people as distinct from the 2,500 that would have been required in a nonautomated factory. The company built this factory not in Japan but in the United States, where all the major car companies still use manual labor and very few robots.

One of the rules in robotics is that one robot = three human workers; a smart robot replaces five or more humans (see Table 13). The basic reason why robots are preferable to humans is that robots raise productivity and reduce costs, two principal goals of business.

## Governors and the Governed

Up to this point in human history, we have been battered by a host of external forces. The development of the management function in modern society, together with all of the interrelated sciences of systems and control (cybernetics), have given rise to several phenomena that can all be related under the umbrella term of "self-governance".

The systems sciences and technologies have given rise, together with many other sciences, to a human species and a society that is increasingly capable of self-governance, of being in control of one's destiny, of not being at the mercy of forces beyond any form of control. Even the weather is increasingly coming under forces of control; this is also true

[41]See Bylinsky, *Fortune* article.

of social problems (poverty, disease), political problems, health problems, communication problems, and so on.

## Personal Robots and the New Servant Class

When slavery was abolished in the civilized world over the last 100 or more years, the idea of slavery did not die with it.[42]

The idea of slavery is based on the idea that there are fundamentally different kinds of people, the governors and the governed. It does not matter what principle is used to make the distinction.[43] What matters is that the distinction, and the idea involved in it, is made and continues to exist.

The upper classes in Britain and the US, and elsewhere as well, make use of butlers and personal servants still, and that is considered a sign of wealth and status, i.e., of being one of the governors, one of the elect. Similarly, many wealthy people have servants or domestic helpers who in effect do all the chores around the house. Some families use their children as personal slaves; many husbands do the same, or used to.

The ideas of slavery and the dream of having a slave, someone to whom one can say, "Let Charlie do it," a man Friday, are ancient and deeply ingrained ideas that have effected much conflict in human history.[44] Now the age of

[42]See John McCarthy, "Mechanical Servants for Mankind," in the Britannica Yearbook of Science and the Future. *Encyclopedia Britannica*, Chicago: William Benton, 1972.

[43]Aristotle made it in terms of Greeks and Barbarians; others made it in terms of Aryan and the "others"; Americans made it in terms of color, black and white; others make it in terms of politics, e.g., Communism where communists are rulers and all others slaves.

[44]The idea of a mechanical slave is ancient. As we saw in Chapter 2, the idea of a servant-robot has taken hold of people's minds and has led a number of inventors in the past to create ingenious machines that could be regarded as servants. Starting with the divine Hephaestus who supposedly had servant girls made of pure gold,

143

robotics gives this idea a source for legitimate fruition. Robots for personal use are relatively cheap right now, and are sure to get even cheaper and more versatile. Within the reach of most American pocketbooks will soon be versatile personal robots, personal slaves that will change many work and general life habits.

## What Should Human Beings Do With Themselves?

Human beings can either do something or nothing. All doing is the expenditure of energy for the sake of some result or product. When one does "nothing," one expends energy merely to stay alive—to breathe, digest, sleep, that is energy is expended automatically or autonomically.

Robots may take over and perform better than human beings, leaving us with little work to do. This is one of the scenarios that has been well-worked over in the media, and in various episodes of TV series like *Star Trek*. Robots, the creations of human beings, take over and act like the "creators" (which is unfortunately Asimov's view of robots, leaving the creators at the mercy of their creations). "At their mercy" can be interpreted in any number of ways. For instance, robots could simply make sure that each and every want and need were met instantaneously, leaving human beings in a trance-like immobility. It is quite possible that a sensate culture with "consumeritis" could fall sway to a situation such as this, but I doubt human beings would allow this. The remainder of this chapter explains why.

### Freedom and Necessity

Human beings, living things, need various things in order to live. These are necessities. Disregarding the common

we move into the reality of the 11th century when Solomon Ibn Gabirol of Valencia is supposed to have created a female robot to do his housework, and in 1557 we hear of a wooden robot that could fetch an ex-emperor's daily bread from the store.

144

meaning of "necessities" (which may for some people include Rolls Royce's), those things are necessary for human life. Air (or gases), water (or liquids), food (or solids), and heat (or plasmic forms of matter) are necessary for life.

On the opposite extreme are those things that are absolutely free and are not required for human life, such as art, literature, and sports.

What appears to be happening as a result of the visionary leadership of a few people is that more and more of the energy that human beings as a whole could expend on obtaining the necessities of life can be replaced by robotic energy, so that human beings become "free" of necessity and rise to the level of truly "free" beings who do nothing out of necessity and everything out of a sense of freedom.

This theme will be explored by considering four types of energy-expenditure that human beings may engage in, two of which are "necessary" and two of which are "free".

## Necessary Actions

There are two sorts of things that human beings need in order to live. One sort has already been talked about in the form of the absolute necessities of life—air, food, water, energy. When we expend energy for these things, we are exerting energy for the production of constantly needed items which are always being consumed.

On the other hand, there are some "necessities" that are not absolutely necessary, given certain conditions, but which are pretty necessary all things being equal. We can do without houses (as any tramp knows), and we can do without clothes some of the time; but these are necessary at least in some climates. When we exert energy for these types of goods, we do so to produce items which can last for more than just a short time and which do not have to be continuously replenished.

The production of these necessities—lugging water,

ploughing land, stoking furnaces, keeping forests alive—are activities that are what we call colloquially "grunt" work. It is tiring, boring, repetitive. If we have robots, let them do these things. Let them be the producers of the necessities of life.

Once that is done, only two things are left—free externally-directed quests called actions, and free internally-directed quests called contemplation.

### Free Action and Contemplation

Once the necessities of life are taken care of by omnipresent robots, then human beings will be free to exert their efforts in non-necessary directions. If one is not intent on being a do-nothing, what actions do not involve necessity?

If necessary effort is that which is exerted to produce items that must be either continuously or regularly replenished, then efforts of a different kind must be directed to the production of items that are *longer-lasting* and more *shareable* than the items produced out of necessity.

Great undertakings, great projects—such as the effort to land on the moon—are longer-lasting and more shareable than is yesterday's meatloaf or four-year-old jeans. Similarly, great inward journeys, those that require the imagination and the powers of attention, the sort of effort that I am calling contemplation, are similarly shareable and longer-lasting than the items produced out of necessity.

My sense, then, is that rather than doing nothing in a robotic age, human beings will have the leisure and the necessary free energy so as to devote themselves to *actions* and *contemplations*. The result is that there will be an increase in pioneering projects, with many more people exploring the Earth, the seas, and the inward reaches of our planet. There will be a great surge in action-directed efforts. Challenges, expeditions, adventures, pioneering, explorations, and probes will be launched on this planet and outer space.

146

Similarly, we can expect inward journeys, imaginative and contemplative journeys that may well usher in new forms of artistic endeavor and of creative originality. Though I grant that there are many good reasons to fear the rise of computers and of robots, there are more good reasons to welcome them.

# Careers in Robotics 7

This chapter describes some of the career paths that presently exist for people interested in robotics, and speculates on some of the careers in the field that will exist in the years to come. The purpose is not to present anything near a comprehensive compendium of actual and possible careers, but to give the reader a sense or feeling for a robotic career.

## Introduction

The idea of a career is an interesting one. A career implies a sense of longevity, that it is something that people do over a long period of time. To have a career is not just to have a job, though the two are confused in the popular mind. One can hold many jobs without ever developing a career out of any of them. The reason is that a career also has the sense of skill or professionalism in the doing of whatever one does in a career. This in turn implies the development by the careerist of skills and aptitudes over a period of years. It implies that one gets better at one's job as time goes by.

At one time, in recent memory, it was thought that someone in a career was in it for life. A lawyer, dentist, teacher, banker, programmer were expected to remain in a career *over the course of an entire life.* People were expected to be stable for life with a career that they began their working lives with. But in the last two decades, another view has come to prevail. It is that while a career is something that one is in for a long time, and in which one hones one's skills to the best of one's ability, the length of time in which one spends in a career has shrunk. It is now recognized and accepted that people will generally have several careers over the course of a lifetime, that developing skills in several areas is a legitimate way to actualize oneself.

How long one spends in a career is a matter of taste, perhaps. Oppenheimer claimed that one should never spend more than ten years in a career because by that time anyone will have wrung themselves dry of creative ideas in that field. After ten years, he suggested, move on. Others may prefer five or twenty years.

For someone interested in robotics, there are many different paths depending on ability and interests. Some of those different paths are described in this chapter.

## Careers In AI

Research into artificial intelligence began with the interest in and development of the theory and application of the computer. Every computer now in existence can do all of the elementary and many of the most advanced mathematical computations with ease.

Mathematical computations are not very interesting in and of themselves. Remember that it was Pascal who started

150

off the calculating machine craze with the development of a machine that could relieve his father of innumerable computations connected with his business. Such computations are among the most boring and most easily computerized areas of our lives.

Making a machine that can add and subtract is to create an artificial intelligence, but one of a low order. What would be interesting is to discover the algorithms (procedures) and electro-mechanisms involved in all the activities that we generally refer to as intelligent behavior. If we could discover those algorithms and develop the mechanisms (and perhaps even the organisms, now that we are in the age of artificial organisms) that are involved in "intelligence," then we will be able to artificially create intelligence, build intelligent machines, develop artificial intelligence.

But what do we mean by intelligence and intelligent machines?

**What Intelligence Is**

With respect to intelligence, there are four general categories:

> intelligence as a state of mind,
>
> intelligence as a way of using a mind,
>
> intelligence as a way of communicating and using states of mind, and
>
> intelligence as a way of relating to one's volitions, emotions, and sensations.

*Intelligence as a state of mind.* As a state of mind, intelligence is a capacity to avoid foolishness, to understand ideas, to be able to think of and in terms of ideas, to be able to reason, to know how to inquire and research, how to assess facts and develop theories, how to evaluate evidence in favor of different hypotheses.

151

To understand ideas, first of all, requires the ability to symbolize, to generalize, to abstract, to deal with concepts and with mental constructs.

To reason requires the ability to deliberate over alternatives and to make logical inferences about the consequences of different courses of action. It requires, in other words, an understanding of the causal (cause and effect) patterns of the universe.

Inquiry is a more complex intelligent behavior that requires the ability to perceive and experience, to hypothesize, to test and to evaluate the results. The assessment of "facts" is required here.

One aim of AI research is to develop mechanisms with the ability to "have" a mind that can have or be in a state of being intelligent.

*Intelligence as a way of using the mind.* As a way of using the mind, intelligence implies a capacious mental attitude, perhaps an ability for creative thought and imagination, and an ability to look backward (past, memory) and forward (foresight and anticipation).

A different way of perceiving and conceiving the AI research into AI is to conceive intelligence as a power to be used in a certain way. To develop AI is, in a way, to develop "beings" with the power of imagination, memory and foresight. Imagination implies the ability to go beyond sensations and what is sensed directly and to "discover" or "create" images of things and events not experienced. Certainly BACON at Carnegie-Mellon University, the intelligence that can infer laws of nature from masses of facts, doesn't have even a rudimentary capacity for imagination as defined.

Similarly, intelligence implies the ability to compare the present with what happened in the past, and what can be expected to happen in the future. Since it takes anywhere from .2 to .4 seconds for an experience to be transferred from short-term to long-term memory, without long-term memory we humans would be limited in our intelligence to only what happens in the present (²⁄₁₀ of a second). The memory feature is something that has been developed considerably by the computer industry. Robots of the future may be expected to access information stored in memory banks, not only those carried by the robot but also memories stored in data banks all over the world.

To anticipate is to be able to infer from what is going on now to what can be expected to happen in the future. This is another power of intelligence that will need to be developed by AI research.

*Intelligence as the communication of ideas.*   The communication of states and uses of mind is also a function of intelligence. To be intelligent is to be able to communicate ideas intelligibly, to be able to learn, to be able to express ideas in linguistic or artistic ways, and to respond to feedback from the environment.

To communicate intelligibly one needs to be able to understand and then to transfer that understanding to another. At the present time, personal robots can speak using up to 150 words, a rudimentary set of words that is insufficient for adequate communication over a whole range of topics. This area of speech recognition and verbal communication needs to be developed considerably if we are to speak of intelligence. The man who had a great impact on the whole development of computing, Alan Turing, argued that he would call a machine "intelligent" if he were able to communicate with it in such a way that he would not know it was not human. Some clever programs showing how machines can communicate interactively and intelligibly with human beings have been developed, the most famous being Weizenbaum's ELIZA program.

Intelligence goes beyond just the ability to communicate intelligibly. It also involves the ability to learn from mistakes (which personal robots have as an ability already), to explore ideas through language and art (which is something that cannot be done now), and the ability to respond to feedback

153

from the environment (which is another power of current robotic mechanisms).

*Intelligence as a way of relating to volitions, emotions and sensations.* Intelligence functions in filtering, interpreting and responding to sensations. Similarly, it functions to evaluate the reasons and drives or emotions that attract or repel someone from or toward a course of action, and it can serve as the director of a course of action once undertaken.

Intelligence in this fourth sense implies the ability to (1) filter and interpret sensations, (2) to evaluate the pros and cons of a course of action, and (3) to be able to select a goal and to plan a course of action to achieve it.

The first and third of these powers is available, in a rudimentary sense, in current robots, while the second, the area of evaluation of alternatives in light of emotional responses and valuations, is something that has not yet been developed even in a rudimentary sense.

These are the areas that AI is concerned with: to develop artificial intelligence that can think and reason, that is creative, that can communicate ideas, and that can interact intelligently to or with respect to those who are emotional, volitional, and have sensory reactions. AI is a "combination" science, an area of applied theoretical science that has as its aim the development of a machine or an artificial organism that can think, create, communicate, sense, feel and act intelligently.[45]

### Examples of Intelligent Machines

As we have already seen, at the present level of AI development, there are machines, or parts of machines, that can perform intelligent functions:

---

[45]See Tom Lonergan and Carl Friedrick, *The VOR*, Hayden: 1983, for a futuristic robot which does quite a few of these things.

154

- There are machines that think.

    BACON can make generalizations and discover hidden patterns.

    There are programs that can play chess almost as well as chess masters can.

    The robot that landed on Mars could perform preprogrammed inquiries, but it had no volition or understanding of its own.

- There are machines that can create.

    The rules of music can and have been programmed and used by machines to "compose" works of music.

    No machine exists, though, that can be said to be imaginative.

- There are machines that can communicate.

    Speech synthesizers have been developed so that human speech can be duplicated by a machine.

    In a rudimentary sense, all computers and robots communicate their states to their users.

    Numerous machines have been and are being developed that can teach, guide, and learn from their experience (as do personal robots already in existence).

- Mechanical sensors have been developed.

    Machines can see, hear, feel, touch, grip and sense their environment, though these machines are still in their primitive stages of development.

    Since sensors in and of themselves are not much help, but only in relation to intelligent capacities to interpret and respond, much work has yet to be done in this area.

- There are machines that feel, that are emotional.

    One loudly proclaimed idea is that machines can't feel, can't be emotional.

    It is predicted that the idea of feeling machines can

be realized in a few years, and with mechanisms that are made for "feeling".

- There are machines that can act, can set a course and follow it.

   These have been developed and are becoming increasingly sophisticated.

   Drive/power mechanisms, arm-wrist end-effectors, and multicapability robots will soon be introduced.[46]

### Careers in AI

Careers devoted to researching the nature of intelligence and in developing mechanisms that can perform intelligently are presently, for the most part, limited to academic careers and careers in the R&D divisions of the government and large corporations, and a few individuals, like Ted Nelson who is working on the "world brain". In the future, however, the research and development of AI can be expected to spread beyond academia and large corporations into the general marketplace.

   I have always found that a good way to begin to understand a field of study or an area of knowledge is to read biographies or autobiographies of people who have worked in the field. This is a method that I would recommend to someone thinking about an academic career in AI. Fortunately, there are many sources of information concerning the people who pioneered and now are leaders in the field of robotics and AI. Names such as Von Neumann, Turing, Wiener, and Shannon can be joined to names of previously mentioned people such as Minsky, Simon, McCarthy, Engelberger and Asimov. A good background source on the lives and theories of the leaders in the information, AI, and, by extension, robotics age is to be found in Pamela McCorduck, *Machines Who Think*, San Francisco: W.H. Freeman, 1979.

[46]The journal *Robotics Age* documents these developments as they develop.

156

Another perhaps equally profitable way to proceed is to obtain any one of the spate of books on AI or related fields, such as bionics, that describe *problems* and mechanical *solutions*. Such books will give one a sense for what is being done and some of the results obtained, as well as a sense, perhaps, of the frontiers of the field and of what still needs to be done.

A third way to gain a sense for a career in the AI area is to obtain a job—any job—in a robotics company, or to seek out and work for a company where automation and robots are likely to be used and where a company has a robot education program for its employees, as does Westinghouse.

## Engineering Careers

The function of engineering is to solve problems of a practical nature, using scientific and intellectual tools. Engineering as a discipline and as a body of professional knowledge requires that the practitioner possess a wide and varied background in mathematics and physical and social sciences, with an emphasis on the applicability of all of these to real, practical problems.

It may be helpful to consider the engineer in the following way. A problem (situation) may be encountered by someone who develops a specific request (called by such names as specifications, problem definition, brief, thesis, and so on) for a solution to the problem. This is phase 1 of the development of a solution.

Phase 2 begins when either the person directly involved— or a design specialist—develops a design solution to the problem; that is, he maps out the logic of a solution.

Phase 3 is when the engineer enters the scene. It is his task to use his knowledge of the materials and construction constraints (e.g., laws prohibiting such and such structures) to solve the problem of how to actualize the design. The engineer thus has to be both versed in theory (mathematics, science) and also knowledgeable about what materials have

157

what powers and strengths, and what constraints there are on using different sorts of materials to build different sorts of constructions.

With regard to careers in robotics, there are different paths than those with an engineering bent of mind can follow; such involve the development of hardware related to robotics, the actual machines, and one directed on the software or procedures involved. This is, of course, a rough distinction since it is becoming harder to distinguish software from hardware; at some points they merge indistinguishably into one another.[47]

## Hardware Engineering

There are several different types of robotic hardware. In each case, the problems and the machines will be different. Three areas of hardware development are presently prominent—

bionic hardware

industrial robotic hardware

personal, multipurpose, robotic hardware.

### Bionic Hardware

Bionic-related hardware engineering seems to be most prevalent in medical circles. Ever since the movie *The Six Million Dollar Man*, and the subsequent TV series of the same name, the idea of bionic people, of machines merged with animals (including and particularly the human animal), has

---

[47]It is worth noting that a Harvard University study has suggested that at the very most, only 30 percent of all available jobs are "advertised". The remaining 70 percent of jobs are "created" when someone walks through the door and company personnel do not see how they can let that person go back out, so "create" a job on the spot. I expect that in robotics, most of the jobs will be created rather than advertised for the foreseeable future.

gained popular acceptability and support. The hero of the movie ended up with two artificial legs, one hand and arm, and an eye, all which possessed greater power than an ordinary human part, and which together gave the hero powers such as leaping 30-foot heights.

The field of bionic research was not named until 1960 when the government sponsored a conference on the topic. A similar phenomenon occurred with AI which appeared in the title of the first AI conference at Dartmouth in 1956.

Throughout history, there have been many bionic-related parts used to replace human organs and tools. False teeth, peg legs, and hooked hands have been popular in both fact and faction. Bionics differs from these in that it requires the collaboration of AI, the collaboration of people working on the nature of intelligent machines. It is developed with the use of sophisticated design and thinking machines. For example, the hand is a remarkable tool. It can swing a hammer, pet a baby, smash a concrete block, arrange flowers. Designing and developing a useful hand by mechanical means does not necessarily involve duplicating the human hand; for instance, a human hand has one thumb opposed to four fingers, whereas a typical personal robot hand has only two digits opposed to each other. Some assembly robots, though, have 20 or more fingers on their hands. In cases where artificial—myoelectric—arms have been created, they are stronger than human arms and handgrips.

Using mechanical or artificial replacements for human organs and limbs, and for the organs and limbs of other creatures, is a highly interesting area of work. In all likelihood, a particular area of work would have to be chosen. In addition to developing arms and hands, there is also work to be done on developing the senses, both those that we have as well as those we do not, so that they are stronger or better than our "natural" ones.

For example, bats use their acute hearing to fly in the dark without hitting any one of their thousands of neighbors;

159

ants use chemical scents to mark trails; fish use taste; a cockroach has the power to detect movement (using touch sensors) of $\frac{1}{254,000,000}$ of an inch; rattlesnakes use heat sensors for differentiation. Such creatures have the senses that we do, but have sense organs which are much more accurate and powerful than are ours. There are also animal senses that we do not have. For example, some fish can sense electricity, which we cannot do very well, if at all.

Developing mechanical organs and limbs that can duplicate and surpass the powers of the senses, and other organs and limbs, is one career path in AI.

There are numerous ideas on bionic-related hardware and software mechanisms. If we look at a human being or other animal, there are three principal features which can be bionized:

1. arms and legs (manipulators and movers)

2. internal organs

3. the brain and other senses located in the head.


Bionic research is concerned with the development of artificial replacements for these three specific regions of human "physical" capabilities. The aim is to replace components in any of these areas should the original part become defective. Myoelectric, bionic, arms-wrist-hand combinations exist and are used to replace lost limbs. Unlike hooks and other mechanical replacements for such lost limbs, bionic parts are created with the aim of duplicating the function of the lost limb, including the ability to have the limb respond to commands from the brain. Similarly, research is being carried out to create artificial replacements for the internal organs. (The artificial heart developed in Utah is one example.) We have already noted some of the areas where research is being carried out to duplicate powers of the brain.

160

*Industrial Robotic Hardware*

A second area of development is in industrial robots, mostly single-function machines that can do one task extremely well, what we can call a dedicated machine or robot. Most industrial robots in existence today are single-purpose or limited-purpose machines. In pure form, these robots possess lifting (e.g., hydraulics) or manipulative powers of great precision and versatility. Arms and wrists (end-effectors) move, twist and position themselves with great accuracy and reliability. Welding or spray-painting robots are such limited-purpose devices.

On the other hand, though single-purpose robots dominate in industry today, there is the realization that the most effective, productive, and affordable use of robots will require the development of general-purpose industrial robots that can be used for a variety of tasks. Those people who enter the robotics field at the present time can be expected to participate in the transition from single-purpose to general-purpose robots.

*Personal Robotic Hardware*

The area in which I myself would be most interested is in the development of hardware for general-purpose personal robots. This area is interesting because it involves combining machines of different sorts (such as machines that see with those that think) so that they occupy as little space as possible, and so that the machine is capable of doing many different actions with its limbs and organs.

While a robot can have any shape—there are robot satellites in space right now—and can be of any size, personal robots will be most useful, I think, if they resemble humans as much as possible. If we want our personal robots to serve as servants and companions, to handle the house the way *we* would handle it, for example, then robots will have to be like us in physical and intelligent functions. This is one reason why personal robots are generally of "our" size and

161

nature. Though they may be humanoid or nonhumanoid, the overall sense one gets is that both the manufacturers and the users want and expect that personal robots will have "personal" characteristics. Personal robots that are expected to be our helpers and companions need to be as much "like us" as possible so they can fit into and complement our lives. They should be able to adapt themselves to our environment, to our way of living. We should not be expected to adapt to robots.

## Software Engineering

Software can be created in the way that engineers proceed, that is, systematically, or it can be done in a haphazard manner. Those who are called upon to create software for computers are increasingly being called upon to systematize their procedures, that is, to become software engineers. Similarly, when we turn to robotics, we will find that systematic software and software engineering will be a requirement for people developing software for robotics. I say this for the following reasons.

There are several levels on which human minds operate, and thereby several levels of software (see Table 14).

At the lowest level of software, we find programs that perform one function. There are programs to compute square roots, print a dot on a screen, make a robot move in one direction. This is the level of elementary tools. A programming language is a set of tools of this nature. Just as a carpenter will have a toolbox of nails and screwdrivers and hammers, so a software engineer must have a set of tools like a programming language suitable for programming a robot for elementary actions, e.g., move forward, move backward, turn left, turn right, and so on. Thus, a program for a robot operation is similar to a program for a purely computerized operation, but the robot program would contain commands related to physical movement in real space and

162

Table 14. Levels of robotic software.

| Level | Nature of Operation | Example |
|-------|---------------------|---------|
| 1 | Elementary operations | Move one step |
| 2 | Sequence of elementary operations | Map out a room |
| 3 | Complex of sequences | Vacuum floor Wash windows Feed the cats |
| 4 | Decision-making | Take care of the house this weekend |

not just movement in virtual space, as on a CRT screen. For example, a typical robot program would contain statements (commands) such as:

```
10  MOVE FORWARD ONE STEP
20  EXTEND ARM
30  BEND ELBOW ONE DEGREE
```

On a second level, we find programming language tools combined into a sequence (called a program) that can be used by a robot to perform a continuous activity. For example, we might program a robot to learn about its environment and to explore it continuously until it needs some food (more electricity). The program could have a few simple procedures:

Start where you are and move forward one step.

If you encounter an obstruction, turn left and go one step.

If you cannot move in any direction because of obstructions, cry out for help or shut yourself down.

If you can move in a direction, keep moving until you

163

encounter an obstruction, then follow the second step above.

All the time you're doing this, store in your memory the places where you have encountered obstructions and "free" movement areas so that you don't have to learn the environment all over again every time.

While you're moving around, investigate any loud sounds or lights.

Software on this level is already developed in its design form, and most personal computers possess "learning" software of this nature as standard software.

Another level of software for robotics is that which combines a knowledge of the environment with a capacity to respond intelligently to it in terms of computational power. That is, a third level of software is achieved when motive power is combined with intelligent power.

A fourth level of software is able to control, monitor, and integrate elementary sensory and motor/drive powers with functional software, such as the ability to map out an environment, and with intelligent computational power to achieve a multipurpose general robot. This is the level of software development that is in greatest need today, and the area that is most promising for short-term development.

This level of software is that which helps us achieve a truly versatile robot. Software at this level would be sophisticated enough so we could simply say to a robot on walking out the door, "Look after the house while I am gone," and it would know what that meant and how to accomplish the many different tasks involved in house-sitting. This is software with a host of (1) elementary operations, (2) many sequences (programs) of elementary operations, (3) complexes of sequences, and (4) autonomous, decision-making software systems (see Table 14).

We are still at levels 1 and 2, in the early stages of level

3 software, but level 4 software is only in the speculative stage.

## User Careers

In addition to theoretical and engineering careers in robotics, there are and will be numerous user-based careers.

There are three types of user careers:

*Robot operator.* As industries and manufacturers install more and more robots in the workplace, those people who neither fear robots or are ignorant of them will be in a position to take charge of and use robots in the workplace. Just as the computer industry at first caused fear on the part of many people, so robots will initially cause fear in the workplace. Those people who wish to get into the robotics field and who are not afraid of these new machines are going to be called upon as operators in the field.

Operators are expected to know *how* robots work, not necessarily *what* they are used to do. It is the task of robot operators to know how to operate robots, to recognize when a robot is not operating properly, and when it needs maintenance and adjustment.

*Robot programmers.* The programming profession arose with the computing industry. This is the profession of those who are able to understand procedures and to translate those procedures into programs that can be understood and executed by robots. Given the myriad of uses to which industrial and personal robots can be put, robot programmers will be busy in the years to come.

*Robot maintenance.* "Behind every successful robot there is a technician."[48] Just as the computer industry spawned the need for technicians to maintain these machines, so robots

[48]*New York Times*, October 17, 1982.

165

A Layman's Introduction to Robotics

will spawn the new career of robot technician and maintenance. By 1990, some analysts predict there will be close to
200,000 industrial robots in the workplace. There will be a
need for people with skills and experience in both electronics
and mechanics who can do preventive and corrective maintenance on robots. In many ways, this area is most suitable
for workers displaced from other fields, such as the auto
industry.

## Education and Jobs

There are two principal ways to get an education needed for
work in robotics. One is through schooling; the other is
through direct experience working with robots in industry.

Unfortunately, academic institutions are usually way
behind everyone else when it comes to topical learning that
is relevant to current problems. For this and other reasons,
academic schooling in robotics is not recommended. At the
current time, there are less than 30 colleges and universities
with full degree programs in robotics. Most of the courses
taught in universities and colleges related to robotics are in
the form of introductory[49] courses. This is a woefully inadequate level of support for the industry. Many academics,
particularly those not in business schools, tend to be wary
of technology and reject new technology as easily as Luddites
do. Academicians will catch up only when robots have become
an inescapably pervasive part of the society.

In the meantime, those people who wish to gain a
schooling in robotics would do best to enter industry and
gain theoretical knowledge as well as practical knowledge of
the actual uses and benefits of robots. The robot revolution
is sweeping the manufacturing industries, and job oppor-

[49]See "Current Offerings in Robotic Education," p. 28–31 of *Robotics Age*, Nov./Dec.,
1983, Vol. 5, No. 6.

tunities available in robotics right now are mainly in manufacturing.

All the automakers are currently considering, preparing for, or actually using robots. Because of the many social and economic problems caused by robots and automation, it can be expected that many US companies will involve their employees in in-house robotic education programs.

There is another form of school which I recommend— self-schooling. In 1976, for example, I was teaching logic and theoretical foundations for doctoral candidates. At the same time, I became interested in computers and was reading every book on the subject that I could find. However, I was quite apprehensive about actually *having* a computer. Then a wonderful event happened: suddenly in 1977 and 1978, the commercial microcomputer appeared in the form of advertisements for affordable machines from the likes of Apple Computers, Sol Technology, The Sorcerer, and several others. Vowing to find out once and for all what computers were all about, I was one of the first people to walk into a Computerland store and buy an Apple II computer. From that day onward I was hooked. A similar event happened in the case of robots.

Book-learning is fine but has little to recommend it for those who want hands-on experience with new technologies. In addition to working in industry, one principal way in which someone can get both book-learning and practical experience with a technology is through self-schooling.

There are at least two vendors who can supply an individual with a self-schooling approach to robotics. For those who are interested primarily in personal robots, the Heath Company provides an extensive course for home-study. If you also purchase a Heath HERO-1 personal robot, you can get both the book learning and the practical hands-on experience that I recommend. The Heath Company is in Benton Harbor, Michigan.

167

# A Layman's Introduction to Robotics

For those interested in industrial robotics, Prep, Inc. supplies a comprehensive robotic training program which includes a well-structured curriculum as well as robotic hardware and software compatible with many personal computers. The address is Prep, Inc., 1007 Whitehead Road Ext., Trenton, NJ 08638.

# Future
## of
# Robotics
# 8

---

## Introduction

As has been true of every invention and advance that breaks new ground, robotics has had its share of detractors as well as supporters. However, since about 1921, robots and robotics have been part of the everyday cultural consciousness, in films, in science fiction, in bionics and prosthetics generally, and in the automation that has increased since the war.

Before considering what future robots and robotics may have, it is necessary to first consider some of the reasons why it is reasonable to support and search out a future for robots.

## Arguments Against Robots

Those persons opposed to robots and robotics seem to fall into three main categories.

# A Layman's Introduction to Robotics

One argument is that there never can be a robot made that could be "better" or even equal to its creator. This idea is misconceived. This is a naive and ineffective argument since it depends for its effectiveness on the capacity of the opponents to persuade everyone to give up the development of robots and of the curiosity upon which their development is based.[50] They simply want to abandon such development altogether.

A second argument is that if robots could be better than their human creators, then robots would/might take over and enslave/eliminate humans, so it is better not to create robots. This argument is based on fear, misunderstanding, and analogizing, but not on ignorance as is the first argument. It is based on the fear that a "created" artifact might gain the power and will to overthrow the creator. It is based on the misunderstanding over the nature of robots, and it is based on an analogy between what has happened to humans when replaced by robots (job loss, loss of sense of worth) and what might probably happen on a large scale when creations turn out to be better than creators. Also involved in this argument is the hidden premise that creations and creators of any kind can only be related in terms of the relation of superordinality and subordinality, of superior and inferior, of free and enslaved, of controller and controllee.

A third argument against robots comes from theological quarters where a variety of arguments coalesce into the idea that to create robots in our image is akin to playing God, a sin at least in the Western religions. To play God is to attempt, like Lucifer, to usurp His position, the Creator. Since usurpers can create only imitations and not real images, the best we humans can create are imitations of ourselves, never images. To attempt to usurp, to attempt to play God, is a sin and should not be done. Of course, these are the same people

---

[50]This is the path taken by Herbert Dreyfus, a perennial critic of AI who in effect says that AI shouldn't be done because it can't be done.

170

who said, "If God had meant us to get to the moon, he would have put us there in the first place," or, "If God had wanted humans to have robots he would have added them to the list of his creations," and other such mind-numbing utterances.

If we were to take these three arguments against robots seriously, there would be no future for robotics.

So, to support the possibility of a future for robotics, let us consider some supportive arguments.

## Arguments in Support of Robots

First, robots can and may very well be better than us in many respects. The different motor/drive systems, manipulators and end-effectors, not to mention the minds that may eventually be built into robots, may very well make them better than we are in many respects. They may be tireless, not bored by repetition, stronger, with better eyes, ears, more sensors, better memories, capable of many more feats of endurance and strength. Nature's tools have already been improved upon by human ingenuity throughout the history of technology. We have built better paths (through roads and bridges), better locomotion (planes, buses, cars, etc.), better food, better lighting, better warmth-provision, better communication systems. Why not build robots that can do many of the very things we cannot do because of our limited powers and fragile physical mechanics? Besides, we will never know whether or not robots can be better than we are until we try.

In opposition to the second argument in the previous section, we have a two-part argument. Cars, tractors, toasters and ovens are much better than we are and have not taken over, except in the sense in which religious zealots warn of slavery to "material" things. Similarly, there is no need to expect with horror the idea of robots taking over because the very idea of conceiving of robots and humans in terms of super- and subordinate relations is not warranted. In fact, the relationship that best characterizes that of human and robot is one of complementarity and parity. Conceived of as helpers and cooperators, robots lose their threatening characteristics.

Against the argument that we should not be playing God, we can, for those who insist on considering robotics under the theological viewpoint, simply say that humans are "co-creators" and that playing God is in fact what we are *supposed* to be doing.[51] Improving on "creation" (if indeed there was one) is in fact the human imperative. If robots constitute one such avenue of development, then they are part of the human task.

These are the reasons that I see as being at the basis of the future of robotics.

## Robots and People

The economies of all advanced industrial nations are in a state of transition to new economic forms. While energy and energy-related products have dominated the economies of the last 100 years or more, information-related products are predicted as being at the basis of the future economies in advanced nations. A sign of this development is the degree to which information-related industries have taken hold in the economies of Japan, Europe, America, and Israel. We are in the midst of a transition from energy to information and from human to automated production.

While I agree that we are in a transition to an information-based economy, I also believe that information has been conceived of too narrowly and that the dominant economy of the coming age is one that will be dominated neither by information-processing (i.e., computer machines only) nor by automated production facilities only, but as one where

---

[51]See, for example, my article, "The philosophy of R. Buckminster Fuller," *International Philosophical Quarterly*, XXII, No. 4, December 1982, pp. 295–314.

the informational and productive, computers and automation, merge into one—in robots.

The robot is, in other words, the machine that will really create a new economy. It is not the computer that will dominate the economy, nor will industrial robots. Rather, it is the robot that combines the powers of industrial robots and computers that will dominate.

In my view, computer-mania has just about run its course. Computers are useful when they can do things. The programmable, multifunction robot is a natural extension of a computer and is what probably justifies having a computer in the first place. While I have the power of some of the largest and most powerful computers in the industry at my disposal, I have never had to use one to balance my checkbook, to entertain me, or to help me track my stocks and bonds.

Once people realize that computers are relatively worthless in and of themselves, and get tired of computations and number-crunching, then they will start looking around for machines that they can connect to their computers that will help them instruct that machine in doing various and sundry things. I already have plans for six personal robots of my own and several more for my family, so I know I am ready to be a multi-robot family man.

## Robots in Our Future

Once it is realized that in order to get computers to *do* anything we will have to hook them up with robots, then the field will be wide open as to what robots can do for us and what we can do with robots. Before the end of the century, I foresee omnipresent robots of all sorts doing many different kinds of jobs in the economy and considerably improving the quality of life for many people.

Consider, for example, old people. Loneliness and boredom, not to mention physical disabilities that prevent per-

173

forming elementary tasks, are often the lot of old people. Robots that could amuse, teach, and do many of the chores that old people find nearly impossible could be made. This is also true for the handicapped. Just as computers were a boon to many handicapped people who through a computer were able to find ways to communicate and create, so robots would extend the powers of handicapped people considerably.

Household robots will do many of the chores around a house, such as vacuuming, cleaning, picking up, garbage collection, watching, playing with pets, and baby-sitting children.

In the information gathering areas alone I expect that there will be a great many robots in use. In oceans and lakes that have not yet been sufficiently explored, we can expect to see robots at work. How many more marine species heretofore unknown to marine science will we find when using robots to sweep the ocean floors from California to Australia, and from Greenland to Antarctica? What of all the flora and fauna on unexplored land that have yet to be discovered, recorded, catalogued? Similarly, in space we can expect to see intelligent robot satellites plying the trades of maintenance workers, diagnosticians, and repairers.

## Social Uses of Robots

A number of social, industrial, business, and planetary uses for robots have already been mentioned throughout this book. They can serve as:

helpers for old and disabled people

researchers and explorers of unknown terrains and seas

personal servants

workers in space, mines and factories

general substitutes for human labor and work.

In industrial situations, robots are being used for spray painting, welding, machine loading, assembly, and general maintenance. In corporate headquarters, for example that of American Cyanamid, robots make the daily rounds delivering mail; at others, such as robot manufacturers, robots themselves are observed and tested. While it is true that in most of these cases the robot is single-purpose and semi-hard-wired, in principle, there is no reason why industrial robots cannot also be general-purpose. Thus, there is no reason not to believe that at some time in the future, American Cyanamid will have a general-purpose robot with arms, brain, sensors, speech capabilities, and general maneuverability involving wheels as well as other means of locomotion. Such a robot would then have the capability of:

locating and picking up mail from all sources

sorting the mail

delivering mail to appropriate recipients

delivering personal, spoken messages

monitoring conditions along the route, fixing normal problems, and seeking help otherwise.

On the shop floor, away from the decision-making offices, robots are even today doing much of the work that people used to do, including loading and unloading machines, maintaining machines, doing assembly work, packaging, crating, painting, welding, and testing products. Increasingly, robots can take over the functions on the shop floor. Manufacturing then becomes a matter of assuring that the proper materials get to the right place at the right time, and not with what shift it is, or with whether or not some workers are ill.

Much of the clerical work done today can also be taken

over by robots—sorting papers, locating information, keeping an appointments schedule, monitoring performance, getting the coffee and doing so many of those "indispensable" chores required in a business, academic, or office environment.

If the work environment is out-of-doors, robots can replace or substitute for human workers in any number of places. Instead of having a live policeman in the middle of an intersection with icy streets, a robot could do just as well without constantly fearing for its life. In other work-environments fraught with danger or hazard, robots could be made to do the work.

The industrial and social uses of robots will not be simply a local, North American or Japanese, phenomenon, but will be world-wide.[52]

By the year 2000, it is expected that robots will be most prevalent in Japan, primarily because the Japanese have a head start given the commitment of their government to the development of robotics. The figures for the number of robots in use in Japanese society by the year 2000 ranges from 12,000,000 to 50,000,000. Mitsubishi is already using robots and plans to have factories almost totally run by robots. In addition, the idea and practice of using personal servants is part of the culture, and just about every Japanese individual is expected to own a practical personal robot by that time, or shortly thereafter.

The United States is expected to come in second in the race with from 7,000,000 to 20,000,000 robots, industrial and personal, by 2000. The automotive, electronics, and space manufacturing industries are expected to be the dominant users of robots. Once personal or home robots with sufficient intelligence to perform some of the more tedious chores

[52]See Robert Weil, Ed., *The OMNI Future Almanac*, New York: World Almanac Publications, 1982, pp. 177–180.

around the home are available, individual citizens can be expected to have their own personal "servants" also.

The USSR is expected to follow Japan and the US in developing robotic mechanisms, except that their emphasis will be quite different. The principal reason for this difference lies in the form of government in the USSR. While Japan and the USA are democracies and thus do not oppose the democratization of computers in the form of personal computers, the USSR is opposed to having ordinary citizens have access to personal computers and all the access to information that that entails. Similarly, we can expect the USSR to oppose the democratization of robots in the form of personal robots. Their emphasis in robotics will be on industrial robotics, if anything.

After the USSR come the European countries generally, followed by the third (industrializing) world and fourth (unindustrialized) world countries.

## Personal Uses of Robots

As I've said before, I already need six robots to do the things I need done, and I'm a simple man with relatively few needs. Think of the normal family with a myriad of needs, and you may be talking about a dozen robots or more for each family— robots to watch inside and outside, to clean the rooms, to cook, to keep the roses blooming, to drive one around, to do shopping chores, to play with the kids, to go on wild adventures . . .

Given sufficient memory storage for long programs, programming skill and patience to develop those programs on the part of people, and robot capabilities to execute complex tasks, all of these robot functions can be performed—on a minimal level, to be sure—by personal robots. Once the intelligence capabilities of robots have been developed further, in a few years additional tasks can be assigned to robots,

177

tasks that require judgement, decision-rules, reasoning capacity and precision in execution.

I see four main areas in which robots can function in personal life:

as a source of amusement

as a method of self-knowledge

as a labor-saving replacement

as an enlarger of personal capabilities.

### Robots as Sources of Amusement

Human beings love to amuse themselves. That is, they love to be entertained, to enjoy themselves, to relax, to have fun. They love revelry. For these they hold feasts, parties, banquets, carnivals, fairs and other gala events. In addition to carousing and debauching, humans find entertainment in sports, gymnastics, regattas, matches, tournaments. There is even an entertainment industry that offers theaters, cabarets, ballrooms, casinos, playgrounds, pools, parks, games, movies. Robots can serve a variety of purposes in this regard.

Robots are already plentiful in films. When robots are plentiful, it is not impossible to imagine that there could well be special robot parties, robot games and competitive matches (my robot can beat your robot), even robot theater, ballrooms, and fairs. Once very powerful and general-purpose robots are available, we can well imagine that robots could take part in sports events and in other games of chance.[53] Robots could certainly be programmed, even perhaps today,

---

[53]Some people might think that if you have robots playing a game like baseball, you would have a very boring game. But as has been found in the case of chess-playing computer programs, some of whose methods could be used in intelligent robots, computer chess players are not completely predictable. The "synergy" created between a group of unpredictable robots would be somewhat like that between human players. In other words, robots can mess up, make errors, and so on.

178

to play backgammon, billiards, catch, checkers, chess, dom-
inoes, horseshoes, shooting and tic-tac-toe, maybe even tug
of war (but they cannot yet be made to perform as humans
do in baseball, climbing, fencing, handball, ice hockey and
other such games).[54]

### Robots and Self-Knowledge

Just watching a robot learn its way around a house is also a
source of much amusement, since a robot can exhibit all the
silly, stupid, clumsy traits to which we[55] humans are prone.
We do seem to get amusement from seeing "ourselves" in
the actions of another. Clumsy initial attempts to program a
robot to be cheeky, witty, cagey, or even deceptive may also
be sources of amusement.

A second area where robots will show their usefulness
follows directly from the previous paragraph. Because robots
are really patterned after ourselves, are attempts to duplicate
our powers and even to improve upon them, robots in action
constitute a clear source of knowledge about ourselves as
human beings. It has been said that the central tendency and
aim of human beings is to become conscious of themselves,
to be aware. The view that awareness, consciousness, and
understanding are key ingredients of the human purpose

[54]A competition is being sponsored by the organizers of the IPRC (the International
Personal Robotics Conference) for 1986 that will have robots playing ping pong.
The first tests in this competition will take place in Albuquerque in 1985 with the
main competition due to take place there in 1986. The objective is to ascertain
whether competitors can develop robots able to "see" a ping pong ball (in all
likelihood the ball will be colored, as white is difficult to "see") and to respond
quickly enough to return a ball over a net. For those who are interested in gaining
more information about this competition, or in actually developing robots to take
part in it, contact Sharon D. Smith, IPRC, 777 Locust Street, Denver, Colorado
80220.
[55]See David L. Heiserman, *Build Your Own Working Robot*, Blue Ridge Summit: Tab
Books, 1976, is an interesting book in this regard. Interspersed throughout his
book—which is filled with electronic and electro-mechanical terms and drawings—
are occasional philosophical and experiential comments on robots, and in particular
on the robot that he is (at least as of that time) continually building and improving.

179

lies at the very basis of much of Western society. The whole scientific tradition is based on the idea of not taking things for granted, of attempting to understand and explain, which one can do only if conscious of the phenomena of nature.

A common observation is that while human beings have explored outer space, we know very little about the planet itself; and while we have researched just about everything external to ourselves, we have not explored and come to know ourselves as a species as well as we could. Adults use their children, in part, as sources of self-knowledge. Robots, since they are themselves "children" in their minimal capabilities, can also be sources of self-knowledge. As their powers increase, so we will be able to have adult counterparts from whom we can learn.

*Robots as Labor-Saving Replacements*
The third area where robots can be of use in personal life is probably the area that will determine the course of robotics most strongly—in performing tasks that must be done, and doing it cheaper and better than anyone else. Right now, only visionaries and technical people can see the use of robots. For the general man-in-the-street, the uses and capabilities have not yet become specific and detailed enough, or prosaic enough. Few people can afford the idle curiosity of watching a robot clumsily learn the layout of a room. They want a vacuum cleaner, window-washer, garbage disposal and general handyman all in one. Even if a robot can do a lot of chores, why have a robot do it? Once that question can be answered, the robot will become as prevalent as the TV set. I have already shown that I think this is a matter of time. Robots can do things that will free me for other purposes.

Alvin Toffler[56] supports this in his idea that the "third wave" economy of post-industrial society has as one of its characteristics the idea that dehumanized and dehumanizing

[56]See *Previews and Premises*, N.Y.: Morrow, 1983, pp. 31–53ff.

180

work will vanish. Whereas industrial society consisted mainly in brutal work, information is being substituted for energy and labor in a new economy. Robots are now on the job in the US, Japan, Sweden, France, Germany and elsewhere, doing such things as welding, painting, grinding, cutting, lathing, drilling, using machines, making cars, and so on. For example, at the Chrysler plant in Newark, Delaware, 98 percent of the welding is done by robots.

*Robots Enlarge Personal Capabilities*
The fourth area in which robots can enhance our personal lives is in performing tasks that we ourselves cannot do. I would like to have a robot that could burrow its way several thousand miles into the center of the earth, and could provide me with pictures and sounds and other readings of its environment at different points. However, the technology for such a robot has not yet been fully developed.

As we move increasingly out into space, we have no choice but to entrust a lot of the work—initially at any rate—to robots with great capabilities. Instead of sacrificing human life unnecessarily in hazardous situations on earth and in outer space, we can use robots that learn from their mistakes, that can explore and learn about their environment and how to deal with it. Robots can make the planets and asteroids habitable even before human beings arrive.

The Viking Mars lander did its work without human intervention. It was able on its own to check its altitude, thrust, position and to descend into an unknown terrain satisfactorily. We need robots that can do such things and much more besides, especially for environments that are quite different from Mars, such as Venus with its hot and deadly atmosphere.[57]

In addition and very importantly, we need robots if we

[57]See Carl Sagan, "In Defense of Robots," Chapter 20 in *Broca's Brain*, NY: Ballantine, 1980.

A Layman's Introduction to Robotics

are to explore much beyond the confines of the Solar System. At the speed of light, it would take 4.3 years to reach the nearest star. At considerably less than the speed of light, it may take us thousands of years to get to our nearest neighboring solar system and around ten billion years to explore the nearest 2,000 star systems. Clearly, no human will be able to endure such distances and time. But robots could do it.

The future of robots on Earth and in outer space appears at this juncture to be an exciting and important development in the powers of human beings. Putting robots to use on and under the earth, in the air, beneath the sea, in dangerous and hazardous situations will be an interesting phenomenon to witness during the coming years. Similarly, putting robots to use in space and other exploratory roles is something that will make practical and feasible the aim that some people on this earth now have, namely to be "universal citizens".

182

# Glossary

**Algorithm**    Originally a mathematical term meaning "a definite procedure" for solving a problem. Appropriated by the computer and robotic fields to mean a logically coherent series of programmable steps to solve a computation problem.

**Android**    An invention of scifi writers. An android is a robot that mirrors the foibles of its creator. It is unlike other robots that supposedly transcend those foibles.

**Artificial Intelligence**    A sub-branch of computing, formally begun at Dartmouth College in 1956. Its aim is the development of artificial counterparts for human and other intelligence. While any computer, including Pascal's machine, is a form of AI, the AI people are more concerned to duplicate the "higher" powers such as thought, judgment, perception, imagination, and so on.

**Automation**    Any device, mechanical or electro-mechanical, that has built-in control systems so that it can control and direct its own motions and activities.

**Bionics** Biologically-based artificial life or organs. The focus of bionics is the replacement of human organs by artificial organs that do a superior job.

**Cartesian Coordinates** Cartesian means "from Rene Descartes". This is a system of coordinates developed by Descartes (1596–1650). The coordinates are the x coordinate along one dimension, the y coordinate along another dimension, and the z coordinate indicating up/down along the third dimension.

**Cartography** The science and art of making maps.

**Controller for Robots** A program that monitors application and other programs in a robot, including the function of monitoring the sensors, energy (electricity), end-effectors, and other activities and functions.

**Cybernetics** Wiener's name for a discipline of knowledge acquisition and evaluation, invented by him, that encompasses many other disciplines such as systems theory, information theory, the theory of automata and principally the theory of control (kybernetike) in humans and animals. It is a generalization and extension of the ideas of information theory, computer science, and others.

**Cyborg** An early name for bionic beings.

**Degrees of Freedom** In geometry, a degree of freedom is a degree along the circumference of a 360 degree circle. As used with respect to robots, a degree of freedom is a dimension along which movement may take place, which means that end-effectors are more capable the more the degrees of freedom they have. The easiest image for a degree of freedom is the ability to move left and right (two degrees), forward and backward (two degrees), and up and down (totalling six degrees).

**Drive/Power System**
The electronic, mechanical, or electro-mechanical components that allow the robot to propel itself from place to place and to do useful things with its end-effectors (e.g., arms).

**Droid**
A droid is a robot that is not evil or not capable of hurting people.

**End-Effector**
A robot as a whole is the end-effector of a computer that directs and controls its actions. Likewise, power/drive systems of wheels and other devices for locomotion are end-effectors of the robot. End-effectors particularly in mind are devices like arms and/or grippers that can manipulate objects.

**Ephemeral-ization**
The "doing more with less" characteristic of the electronics industry, where ever-smaller chips with ever-greater power and at ever-decreasing costs are being created. The term is Buckminster Fuller's.

**Freedom**
The possibility of movement in space without constraints. Measured in degrees or dimensions of freedom.

**Governors**
A governor is a technical term invented by Watt (in 1855) to denote some mechanism that directs the behavior of other mechanisms, just

as human governors direct the behavior of people.

**Gripper**   A mechanical device for holding, releasing, and perhaps also manipulating an object.

**Humanoid**   In anthropology, this refers either to prehistoric humans or to human-like animals. Humanoid robots are robots that are human-like, in appearance and/or in capability.

**Information Society**   A society based on an economy of generating and computing information, particularly computerized information.

**Logic**   The study of the formal relations between words, sentences, and other symbolic forms.

**Management Science**   A social science deriving its tools from the physical sciences (e.g., computer science) as well as from the humanities (e.g., common sense), and with a heavy dose of operations research. Its aim is to develop controls and techniques for the mathematical management of events and situations where managers are needed.

**Mathematical Logic**   Also known as symbolic logic. A logic distinguished by the use of variables and relational symbols, much like formal mathematics. The development of symbolic logic can be traced back to the 19th-century work of Boole, Demorgan and Hamilton.

**Manipulator**   A device similar in function to a human hand that can be used to accomplish tasks.

**Operator**   A person who knows how robots work, how they run, and who can initiate and terminate actions by a robot. Similar to a computer operator.

**Production Control System**

Most usually, a computer system (set of programs) designed to provide decision support for manufacturers. A typical production control system provides computer programs or techniques for top management planning, for forecasting customer demand, for monitoring and controlling inventory investment, for distributing finished goals to competing distribution locations, for monitoring production on the factory floor in terms of scheduled dates for finished goods, and for supplying the raw materials and other components that make up the final assemblies.

**Programmable Robot**

A robot that can be programmed and reprogrammed to perform a variety of tasks.

**Quantification**

Putting phenomena into patterns and variables that can be measured and calculated.

**Reliability**

The key characteristic and principal aim of automated production systems and of robotic engineering. Unreliability in the workplace and among suppliers and vendors causes manufacturers today to hold "safety stocks" in inventory so that unanticipated delays or demands can be handled. The aim of automated production is to eliminate the need for safety stocks and other ways of hedging one's bets against unpredictable events.

**Systems Theory**

The theory concerning the nature of systems, and a set of tools for applying the idea of systems to practical problems. The key intuition of ST is that a system is a set of internally

187

related components that affect and are affected by each other in such a way that the whole system is greater than the sum of its parts. The sum of the parts that make up a non-electronic watch is what one finds strewn over the floor after a child is through taking it apart; the watch itself is more than the sum of the parts—it is a system.

**Semantics**
The study of the meanings, connotative and denotative, of any symbols used in communication.

**Sensors**
Artificial devices able to detect events within the range of human sensory organs and also those beyond human ranges. An entire industry is devoted to the development of sensors for sight (include invisible light on either side of the visible spectrum, e.g., infrared and ultraviolet), sound, touch, taste, and smell.

**Servo-Mechanism**
A device that controls the behavior of a mechanism and that is able to maintain that mechanism on a desired course by responding to electrical or mechanical impulses. For example, a thermostat that controls the behavior of heaters and air conditioners by registering temperatures and triggering action when the temperature goes above or dips below a preset level.

**Teaching Pendant**
A common way of teaching robots what to do is to 'show' them what to do using a device that can direct the behavior of the robot without explicit spoken or written (e.g., in a program) commands. This is the easiest method of instructing robots, but also the most limited and simplistic.

**Teleoper-**
**ated**
**Machine**

A machine controlled from the "outside". Useful for hazardous work. This machine differs from a robot in that a robot operates itself.

**Uniformity**

Characteristic of robot behavior.

**Vision**
**System**

This refers to a combination of a mechanical capability, such as a TV camera or other optical sensor, with a computer device used to monitor and interpret receptions.

**Waves**

The wave is one of the most powerful and widespread analogy-models used in the sciences and in ordinary work. We speak of lightwaves, airwaves, brainwaves, radiowaves, ocean waves, and so on. Most if not all sciences that use the concept of "waves" as a way of characterizing a phenomenon rely on the work of Fourier.

# Selected
# Bibliography
# of Robotics

The premier journal is *Robotics Age, The Journal of Intelligent Machines*, started
in 1979. Address: Strand Building, 174 Concord Street, Peterborough,
NH, 03458. $24.00 per volume.

In April, 1984, the journal *Personal Robotics Magazine* was announced by
KLH Publishing Company, P.O. Box 421, Rheem Valley, California
94570.

Robotics International of SME, One SME Drive, P.O. Box 930, Dearborn,
Michigan 48128. RI is a subarm of the Society for Manufacturing Engineers. Dedicated to educating people in robotics and to certifying
professionals in the field.

ASIMOV, ISAAC, *I Robot*, New York: Fawcett, 1970. (Gnome Press, 1950).
New edition of 1950 book where Asimov states the three laws of
robotics.

BERTALANFFY, LUDWIG VON, *Robots, Man and Minds*, NY: Braziller, 1967.
Von Bertalanffy is the inventor of the General Systems Theory, now
widely used in many different fields. Here he applies GST to problems
of human nature.

BYLINSKY, GENE, "The Race to the Automatic Factory," *Fortune*, Vol. 107,
No. 4, Feb. 21, 1983, pp. 52–66.

# A Layman's Introduction to Robotics

CALLAHAN, J. MICHAEL, "The State of Industrial Robots," *BYTE*, 7, No. 10, Oct. 1982, pp. 128–142. Survey article.

CAPEK, KAREL, *R.U.R.*, Garden City: Doubleday, 1923. English version of the 1921 play where "robot" was introduced as a term. It is a Czech term meaning slave or worker.

COHEN, J., *Human Robots in Myth and Science*, London: Allen and Unwin, 1966. Good scholarly source for information on robots in myth and legend.

CULBERTSON, JAMES T., *The Minds of Robots: Sense Data, Memory Images, and Behavior in Conscious Automata*, Urbana, Ill: University of Illinois, 1963. Very technical book by a pioneer in the field.

DODD, GEORGE G., AND LOTHAR ROSSOL, *Computer Vision and Sensor-Based Robots*, New York: Plenum Press, 1979. A technical reference work on research in the field of sensor-based robots.

DREYFUS, HERBERT, *What Computers Can't Do: A Critique of Artificial Intelligence*, NY: Harper, 1972. Dreyfus has made a career of being a gadfly to the AI community at MIT, where he began his teaching career, and throughout the country generally. While many of his views are factually and theoretically unwarranted, he has played a somewhat beneficial role in defending people from being reduced to machines, and by claiming that there is "more" to persons than simply mechanics. His is the protest of the existential person against the seductive charms of mechanization.

GEDULD, HARRY M., AND RONALD GOTTESMAN, Editors, *Robots, Robots, Robots*, New York: Graphic Arts Society, 1979. Excellent source book on robotics for the general reader, with a spate of articles representing many different perspectives.

GLOESS, PAUL, *Understanding Artificial Intelligence*, Alfred, 1979. This book is an overview of the field with introductions to the ideas of AI and its major applications.

HALBERSTAM, DAVID, "Robots Enter our Lives" *Parade Magazine*, April 10, 1983, pp. 18–22.

HARTLEY, JOHN, *Robotics at Work*, Elsevier, 1983. Describes the application and applicability of robots in production.

Heiserman, David L., *Building Your Own Working Robot*, Blue Ridge Summit: Tab Books, 1976. A brilliant and humorous man describes the technical and philosophical issues involved with robotics.

HELMERS, CARL T., *Robotics Age: In The Beginning*, Hayden: 1983. A collection of articles from *Robotics Age* (1979–1981). Good source of information on sensors and speech applications.

HYDE, MARGARET O., *Computers That Think? The Search for Artificial Intelli-*

*gence*, Hilside: Enslow, 1982. A good, semi-scholarly introduction to AI.

JACKER, CORINNE, *Man, Memory and Machines, An Introduction to Cybernetics*, N.Y.: Macmillan, 1964. Sound but narrow.

JACKSON, PHILIP C., *Introduction to AI*, NY.: Petrocelli Books, 1975.

LONGERGAN, TOM AND CARL FRIEDRICH, *The VOR (Volitionally Operant Robot)*, Hayden: 1983. A description and consideration of a design for a robot of the future with high AI capabilities, eyes, speech, and capable of interacting with other robots and with humans.

MCCORDUCK, PAMELA, *Machines Who Think*, San Francisco: W.H. Freeman, 1979. Interesting and noncomprehensive treatment of the field.

MCCORMICK, LYNDE, "US Robots finally giving Factories a hand," *The Rocky Mountain News*, Tues., March 1, 1983, pp. 1B–16B.

MINSKY, MARVIN, *Semantic Information Processing*, Cambridge: MIT, 1968. Minsky's ideas on AI. His forecasts of future wonders of AI seem to irk Dreyfus the most.

NEUMANN, JOHN VON, AND ARTHUR W. BURKS, *Theory of Self-Reproducing Automata*, Urbana, Ill.: University of Illinois, 1966. If you enjoy heavy theoretical works.

O'NEILL, GERARD K., *2081 A Hopeful View of the Human Future*, New York: Simon and Schuster, 1981. One of the country's leading scientists presents an optimistic view of the human future. Worth reading just for the ideas on space.

PATTIS, RICHARD E., *Karel the Robot*, N.Y.: Wiley, 1981. A professor tries to teach robot programming using Pascal, of all things. Nevertheless, interesting and suitable for beginners.

RUBY, DANIEL J. "Computerized Personal Robots," *Popular Science*, May, 1983, pp. 98ff.

SERVAN-SCHREIBER, JEAN-JACQUES, *The World Challenge*, New York: Simon and Schuster, 1981. Tr. from French edition, 1980. The Japanese challenge the rest of the world.

SIMON, HERBERT A., *The Sciences of the Artificial*, Cambridge: MIT, 1969. Good book by a leading General Systems thinker, and one of the original AI workers.

WIENER, NORBERT, *God and Golem, Inc.* Cambridge: MIT, 1964. Speculations of cybernetics applied to religious and ethical problems raised by artificial creations.

————, *The Human Use of Human Beings: Cybernetics and Society*, Garden City, NY.: Doubleday, 1954.

# Selected
# General
# Bibliography

ASIMOV, ISAAC. *Seventy-One Glimpses of the Future*, Boston: Houghton-Mifflin, 1981. Many different facets of the future are covered, but one central theme is robotics.

BATESON, GREGORY, *Mind and Nature*, New York: E.P. Dutton, 1979. Author of the influential *The Ecology of Mind*. Presents his view of the world and science in this, his last (and posthumous) book. An influential thinker.

BERLINKSY, DAVID., *On Systems Analysis*, Cambridge: MIT, 1976. A useful introduction.

BRONOWSKI, J., *Science and Human Values*, New York: Harper, 1965. The well-known popularizer presents his views on the human side of science.

CHURCHMAN, C. WEST, *The Systems Approach*, New York: Dell, Delta, 1968. Lucid introduction to systems theory by the best mind in management today.

DOWNS, R. M., AND D. STEA, EDS., *Image and Environment*, Chicago: Aldine, 1973. How ideas about life and the environment affect both, and how both can be changed.

FULLER, R. BUCKMINSTER, *Synergetics*, New York: MacMillan, 1975. The

195

planet's friendly genius presents his "natural" geometry, the ideas underlying all of his work (for instance, on domes).

LOEB, ARTHUR, *Space Structures, Their Harmony and Counterpoint*, Reading, MA: Addison-Wesley, 1976. A follower of Fuller presents crisp descriptions of space and its structures.

MOWSHOWITZ, ABBE, *Human Choice and Computers*, 2. New York: North-Holland, 1980.

O'NEILL, GERARD K., *2081 A Hopeful View of the Human Future*, New York: Simon and Schuster, 1981. An optimist and a leading creative scientist tells us what he expects.

ROBINSON, ARTHUR H., AND BARBARA BARTZ PETCHENIK, *The Nature of Maps, Essays Towards Understanding Maps and Mapping*, Chicago: University of Chicago, 1976. The best book I know of on cartography with a great deal of relevance to AI and robotics.

# Index
## of Names

# Index of Names

Cohen, J., 16 n, 192
Coleco, Inc., 141
CRAY-1, 12
Culbertson, James T., 192
Cryptozoology, 138 n

Daedalus, 37
Dartmouth College, 159, 183
DaVinci, Leonardo, 42, 44, 47
DeMorgan, Auguste, 47, 186
Dendral, 18
Dertouzos, Michael, xvii n
Devol, George, 53
Descartes, Rene, 30–31, 34–35, 42, 189
Dodd, George, 192
Downs, R. M., 195
Dreyfus, Herbert, 39, 170 n, 192
Droz, 43–44

Edison, Thomas, 43–44
Einstein, Albert, 33, 69
Empedocles, 38
Engleberger, Joseph, 53, 84 n, 131 n, 156

Feigenbaum, Herbert, 16 n
Foulkes, Fred K., xxv n
Fourier, 33–34
Friedrick, Carl, 154 n, 193
Fuller, Buckminster, 22, 41, 172 n, 185, 195

Galatea, 37
General Electric, 125
Geduld, Harry M., 36 n, 192
GENUS, 126
Giarcia, Steve, 61
Gloess, Paul, 16, 192
General Motors, 53, 128, 140
Goethe, 38–39
Gottesman, Ronald, 36 n

HAL, 59
Halberstam, David, 192
Hall, Robert W., 84 n
Hamilton, William, 186
Hartley, John, 192
Harvard Associates, 108, 115
Heath Company, 53, 60, 96, 105, 115, 167
Heiserman, David, ix, 179 n, 192
Helmers, Carl, 192
Hephaestus, 37, 143 n
HERO-1, 53, 60, 65, 96–98, 105, 110, 111, 123, 167
Hero of Alexandria, 41
HUBOT, 98
Hursch, Jeffrey, xxv n
Huxley, Thomas, 133
Huygens, Christian, 44
Hyde, Margaret O., 192

IBM, 17, 49, 50, 85, 90, 118, 124, 125
Ibn Gabirol, 144 n
Iowa Precision Robotics, 97

Jacker, Corinne, 193
Jackson, Philip, 193

Kemeny, John G., 9
Kepler, 72

LaMettrie, 35
Leibniz, xxv n, 47
Lockheed, 17 n
Loeb, Arthur, 196
Loew, Rabbi, 38–39
Longerman, Tom, 154, 193
Lovelace, Lady, 47
Luddites, 38–40, 78
Lull, Ramon, 47

Macsyma, 18

198

# Index of Names

199

# Index of Names

# Index
# of Ideas

*Ideas:* This index is composed of concepts, plans and thoughts of some generality and pervasiveness.

201

# Index of Ideas

202

# Index
## of Subjects

# Index of Subject

# Index of Subject

# Index of Subject

Industry and robotics, 10, 100
Industrial accidents, 114
Industrial economy, xvii–xix, 180–181
Industrial waves, xxvii
Inferring laws, 152
Information, xvii–xx
  high-tech economy, xx
  information-based economy, xvii–xviii
  products and effects, xvii–xix
  society, 137, 186
  theory, 48–49
  weightless, xix
  work, xix
Inquiry, 152
Intelligence, 151–154
  communication, 153
  relational, 154
  robots, 12, 21
  satellites, 21
  state of mind, 151–152
  way of using mind, 152–153
Inversion of God and robot, 132–133
Inward turn, 77
Isomorph, 94

Journals, 191

Key idea of science, 34
Knowledge base of robotics, 11

Law of large numbers, 33
Laws of robotics, 119, 131–134
Learning from mistakes, 153
Leisure, xx
Levels of software, 118–119, 163
Levitation, 78
Light spectrum, 63–64, 69–70
Logic, 18, 47, 48, 186
Luddites, 38, 78, 166

Machine as human, 35

Machine-human interaction, 136
Machine, new species of, 9
Maintenance of robots, 165–166
Management, 136, 186, 195
Manipulators, 9–11, 125, 186
Manufacturing, xxii, 52, 84, 135, 191
  profit level, 140
  robot use, xx–xxi, 140–141, 166–167
  robotics, 85–87
Mapping robot, 120
Mathematical logic, 32, 48, 186
Mechanical
  actions, 41
  duck, 42
  ingenuity, 44 ff
  plays, 41
  servants, 45
Memory, 57–58, 68–71, 123, 152–153
Meta-control software, 118–119, 131–134
Method
  of definite finitude, 120
  of indefinite infinity, 120
  of software development, 121
Military radar and sensors, 93
Minds and robots, 11, 29, 35, 117, 192, 195
Miniaturization, 51
Model of robot, 62
Moral principles for robots, 131–134, 193
Myoelectric arms, 159–160
Mythical robots, 29–30, 36, 38–39, 192

Natural geometry, 196
Necessities, 144–145
Networks, economy of computer, 139

Old people and robots, 173–174
Operators, 165, 186
Operating systems, 98, 118
Order in nature, 31–32

206

Index of Subject

Parallel processors, 72
Party mode robot, 106
Pattern detection, 72
Patterning and procedures, 33
Perception, 31, 65–67, 93
Personality modules, 106
Personal robots, 53, 96, 126
Phases in problem-solution, 157
Philosopher's stone, 39
Philosophy, 30
Pick and place task, 104, 111, 128–129
Playing god, 170–171
Pneumatic drives, 87
Positronic brains, 131
Power-drive systems, 87, 94–95
Prelinguistic programming, 121–122
Prison guard robot, 100
Pro and con robots, 169 ff
Problem-solving research, 17
Procedures, 119
Productivity, xxviii
Production steps, xxi
Professionalism, 150
Programming, 71, 140, 162–164
Programmability, 10–11
Programmable robot, 187
Programmers, 117–119, 165
Predictions about robots, 99, 176
Profit, xxi–xxiii
Promotional robot, 103
Psychokinesis, 78
Pulley and screw, 41

Quality in manufacturing, xxi

Reliability, 187
Religion, 132
Remote-controlled keyboard, 105
Reproducible robot, 51
Research robot, 137–138
Retooling, 99
Revolution, xix, 8, 50
Rights of robots, 79

Robot, 3
action needs, 57
as complement of human, 172
as enslaving, 170–171
as explorers, 174, 182
as general term, 28
as helper and companion, xxv, 162
as isomorph, 28
as knowledge and industry, 1
as labor saver, xxiv, 180–181
as mailman, 175
as manipulator, 3
as melding technologies, 8
as new species of machine, 9
as noun and verb, 1–2
as persons, 6, 79
as probe, 138
as source of amusement, 178–179
as source of self-knowledge, 179–180
as social system, xxvi
as zombie, 6
Robot-Human replacement costs, 142
Robotics
context of, xiii, xv
currently, xii–xiii
educational, 167
information, xii
intelligent, 9
natural outgrowth of computers, xx
new world economics, xi–xii
space necessity, xii
technologies in, xiii
Robotizing, 2–3
Robots
control language, 110
democratization of, xiv
human scale in, 23
manufacturing, xix
widows, 137
working, xxvi, 192

Satellites, xviii, 63

207

# Index of Subject

The intention of this book is to provide a generally educated but nontechnical reader with a continuous, useable book that is not only informative, but also thoughtful, provocative, and practical.

The subject matter is robotics. At the present time, the technology of robotics is about at the stage where personal (or micro) computers were in 1976 or 1977; that is, robots are beginning to enter the market as practical, labor-saving devices. It is expected that the robotics revolution will be just as pervasive as the computer revolution, and that in a few short years, knowledge of robotics will be as necessary for economic survival as a knowledge of computers is today.

This book is an attempt to describe in a general way some central features of robotics, which is one of the principal players in the drama of the "emerging" post-industrial society to which we are all witnesses. This emerging society and economic system is based on *information* (facts, knowledge, data processing) technologies that augment our mental power using tools like computers, and on the emergence of *outer space* as a central feature of our vision of the future. Robots are the chief means whereby the information society will be realized. Robots are the pathway to the automated factories of tomorrow, to the use and development of information as a way of life, and to the exploration and economic exploitation of outer space.

Robots and robotics are relatively new phenomena for the general populace. Most everyone has heard, seen or read something about robots, but robots are still regarded as something "strange out there." Here we wish to deal with robots in context, to dig out some of their salient features, and to describe their implications for human life in general.

First, we need to see robotics in a context. Robotics does not stand alone, but is supported by many other disciplines, professions, and fields of interest. Robotics is supported